The STUNNING SCIENCE of EVERYTHING

NICK ARNOLD ✳ **TONY DE SAULLES**

This book is written for YOU, the reader…

If science feels like a sad closed book
Then open these pages and take a look
Read every word and you're sure to be
A horrible scientist – just like me!
Nick Arnold

Thanks to our creative team:
editors Lisa Edwards and Jill Sawyer
and designer, Michelle Strong –
truly *Horrible* people.
Tony De Saulles

Scholastic Children's Books,
Commonwealth House, 1–19 New Oxford Street,
London WC1A 1NU, UK

A division of Scholastic Ltd
London ~ New York ~ Toronto ~ Sydney ~ Auckland
Mexico City ~ New Delhi ~ Hong Kong

Published in the UK by Scholastic Ltd, 2005

Text copyright © Nick Arnold, 2005
Illustrations copyright © Tony De Saulles, 2005

ISBN 0 439 95905 5

Printed and bound by Tien Wah Press Pte. Ltd, Singapore

2 4 6 8 10 9 7 5 3 1

The right of Nick Arnold and Tony De Saulles to be identified as
the author and illustrator of this work respectively has been asserted by
them in accordance with the Copyright, Designs and Patents Act, 1988.

CONTENTS

THE STUNNING START

Science is about everything. It's about...

SPIDERS AND STARS... RABBITS AND ROCKETS... VENUS FLYTRAPS AND THE PLANET VENUS

And there's much, much more. But that's the problem. There's too much science! Too much to learn, too much to remember. I mean, how can any normal person ever make sense of science?

Well, you could try building a super-powerful computer, program it with every science fact in the universe and plug it into your brain. But this might have unpleasant side effects...

ARRRGH MY BRAIN'S BLOWN UP!

SPLURP!

Or you could read this book... *The Stunning Science of Everything* tackles science in a new way. It's called size-sorting – which means looking at everything in order of size. That's why Chapter 1 starts with the universe in its first second, when it was the tiniest thing ever. And the topics just get bigger and BIGGER until at last you get to see the BIG PICTURE OF EVERYTHING and the whole of science makes stunning sense! (The big picture's on page 92, but don't go peeking at it until you've read the rest of this book!)

We'll be exploring the science of everything with the aid of a crazy collection of made-up characters...

And because this is a Horrible Science book, every page is oozing with horribly funny bits that are guaranteed to make you gag and giggle. But that's not all – since this is a very special posh book, it's printed in gloriously gory colour so you can read about the bloody bits and *see* them too! But don't take my word for it – just turn the page to find out how everything began, and discover why you're actually as old as your granny…

THE BRAIN-BOGGLING BIG BANG

Everything started with a Big Bang. That's how scientists describe the moment the universe popped out of nothing 13.7 billion years ago. But what was nothing like? Well here's a brain-bending experiment to show you the answer...

A thought experiment is an experiment that scientists imagine in their heads. All you need to try one is...

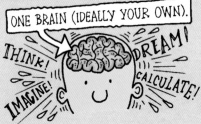

ONE BRAIN (IDEALLY YOUR OWN).

THINK! DREAM! IMAGINE! CALCULATE!

WARNING! These experiments are guaranteed not to make a mess, but don't think too hard or brain juice might squirt from your ears...

SQURP!

Stunning thought experiment – lots of fuss about nothing...

Imagine you're a magician and you make Mr Fluffy vanish. That's right! You pop him into a top hat, wave your magic wand and ABRACADABRA, hey presto – the hat is empty! There's nothing in it!

Now imagine a harder trick. This time you stuff the universe into your hat and make that disappear too!

WOW – now there's NOTHING around you! There's no light and no dark, no space and no time. In fact there's nothing at all. Er, hold on – there's no air to breathe, so you'd better bring the universe back pretty quick!

STAGE 1

HEY- YOU'LL GIVE ME A NOSE BLEED!

STAGE 2

GASP!

Nothing sounded a bit scary, didn't it? I bet you wouldn't want to spend the weekend there. But before the universe there really was nothing at all.

SO WHY DID THE UNIVERSE BEGIN?

BEWARE! You should NEVER ask a scientist this question! If you do, they might gibber and froth at the mouth or, even worse, they might start mumbling about virtual particles in quantum foam and things that happened in other dimensions… The truth is, no one's too sure *why* it happened. All I can say is that the Big Bang would have been BIG news – if there'd been any newspapers to report it…

Mind you, the Big Bang wasn't actually a bang at all. There was no air to carry the sound so it wasn't as loud as your dad snoring. Oh dear, I hope you're not too disappointed…

THE VERY FIRST SECOND

The very first second of existence was a busy time for the universe. In that all-important instant the new-born universe grew from a tiny dot to a fireball billions of kilometres across.

An embarrassing apology from the author...
OOPS, sorry readers – there's been a major malfunction with the size-sorting process! This chapter is meant to be about the most tiny thing ever, but it's suddenly got STUNNINGLY BIG!

Why matter matters...

In that frantic first second, the matter that makes up the universe formed. So that's why I said you're as old as Granny – because the matter that makes you up is exactly the same age.

Matter is what something is made of.

Energy is the power to make something happen. Heat is a form of energy.

Scientists reckon matter formed from energy, so we sent the shrinking scientists back through time to check it out...

The shrinking scientists ... start with a BANG!

THE FIRST SECOND

I'M SO HOT!

WELL, IT IS BILLIONS OF DEGREES.

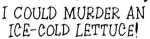
I COULD MURDER AN ICE-COLD LETTUCE!

THE TEMPERATURE IS DROPPING AS THE UNIVERSE GETS BIGGER...

AS THE UNIVERSE COOLS, ENERGY TURNS INTO MATTER.

I WISH THEY'D STOP RABBITING!

YEOW!

WHAT'S THE MATTER?

NEVER MIND **MATTER** – MY TOES ARE MELTING!

The idea of making matter from energy might sound weird, but it happens. Really AWESOME explosions make tiny amounts of new matter and stars shine by turning matter back into energy (as you can find out on page 15). Of course, the person who figured this out must have been a scientific superstar.

Brainy Boffin

Albert Einstein (1879–1955)

Awesome Albert said he was amazed that lessons in schools hadn't destroyed the spirit of curiosity in children. Not surprisingly, Albert dropped out of school and failed his exams for college.

Failing exams doesn't make you a genius, but Einstein was a genuine genius because in 1905 he dreamt up the Special Theory of Relativity. This dramatic discovery showed how energy and matter can turn into each other. And if that wasn't amazing enough, ten years later Albert came up with the General Theory of Relativity to explain how gravity works in the universe. Not bad for a school dropout!

I DON'T CARE IF IT IS YOUR "GENERAL THEORY OF STUFF", JENKINS – ONE PLUS ONE DOES NOT EQUAL XYZ!

FUME!

MYSTERIOUS MATTER FACTS

1 Matter is like ice cream. Ice cream is frozen cream and matter is like frozen cooled-down energy.
2 You need loads of energy to make a pathetic amount of matter. For example, you'd have to blow up a million of the world's most powerful bombs to make enough matter for one peanut. And I guess it would have to be a roasted *peanut.*
3 There's a lot of energy locked up in matter. If all the matter in a new-born baby was turned into energy, it could run a power station for a year.

WHAAAAAA!

ONLY A YEAR?

So it was that all the matter that makes up you, your granny, Earth, and everything else (including the purple slobwobbler from Planet Wibble) was formed in one stunning second 13.7 billion years ago. But that's not all – in that ever-so important first second, the forces that control the universe came into being. And they're so vital that … oh dang – I've just run out of space! To find out about forces you'll have to *force* your fingers to turn the page…

PL'ARF ICK SHNOG FLEF PA YONX DIBBLE? PLURTZ SLOB-TRIX!*

TIZ ICK GOB, GUFFLE ECK DIBBLE?**

* SO I'M THE SAME AGE AS YOUR GRANNY? THAT'S SLOB-TASTIC!
** BY THE WAY, WHAT'S A GRANNY?

FEARSOME FORCES

As I was saying, forces hold together everything – from stars to tiny bits of matter.
And if you've ever fallen on your bum you can blame forces too.

Mass is the amount of matter that makes up something.

An **atom** is a tiny ball of matter so teeny-weeny that you'd need more than 20 million to stretch 1 cm. (You can find out their tiny little secrets on page 12.)

Got all that? Great! The shrinking scientists and Mr Fluffy are safely back in the present, and they've shrunk down to the size of atoms to show you how forces work…

The shrinking scientists … get force-full!

1 GRAVITY

Gravity is a force that pulls things together. The Sun's gravity stops Earth whizzing into space and the Earth's gravity stops you whizzing off into space. And gravity works at the level of atoms…

MY ATOM'S GRAVITY PULLS ON YOUR ATOM.

AND MINE PULLS ON YOURS!

But the gravity of an atom is pathetically weak…

YOU NEED BILLIONS OF ATOMS IN A PLANET BEFORE YOU GET REAL PULLING POWER.

WHAT ABOUT A GIANT CARROT?

The shrinking scientists have shrunk even smaller to study the strong force…

THE STRONG FORCE STICKS TOGETHER BITS OF MATTER AT THE HEART OF AN ATOM.

IT'S REALLY STRONG!

I KNOW – I'M STUCK!

THIS IS A STICKY SITUATION!

2 THE STRONG FORCE

THE WEAK FORCE MAKES SOME ATOMS FALL TO BITS. THIS IS CALLED RADIOACTIVITY.

I THOUGHT IT WAS WEAK!

I SHOULD HAVE STAYED IN MY HUTCH!

3 THE WEAK FORCE

AND IF YOU FANCY A LITTLE MORE LIGHT READING, YOU'LL TAKE A SHINE TO PAGE 18!

THREE FORCEFUL FORCE FACTS TO FASCINATE YOUR FRIENDS

1 The strong force got its name because it's, er, strong. To be exact, it's 100 billion billion billion billion times stronger than gravity. Fortunately, the strong force only works inside atoms, otherwise we'd all be glued together.

2 And get this – the weak force is actually weaker than the strong force! Mind you, it's still ten million billion billion times stronger than gravity.

3 The pull of Earth's gravity on you is 78,000 billion billion times stronger than your pull on the Earth. And that's why you don't pull the Earth up and down when you do press-ups. Of course, the first scientist to explain gravity was a really brainy boffin…

Brainy Boffin

Isaac Newton
(1642-1727)

Isaac Newton must be the brainiest boffin ever. In fact he was so clever that he often forgot to get up, and would sit on his bed thinking deep scientific thoughts. (Don't get any ideas now – that excuse WON'T work for you on a Monday morning!)

Among many other things, Newton worked out a mathematical formula for gravity, which showed…

• The more mass an object has, the stronger its gravity.

• If you double the distance between two objects, their gravity gets four times weaker.

He was dead right, but the most amazing thing was that Newton worked it all out 20 years before he told anyone about it. And when he was asked about it, he'd lost his calculations. Luckily he redid them in a brilliant book full of maths that only other brainy boffins could understand.

Just now we were getting up close and personal with atoms, and that's part of the plan. You see, the next chapter's all about atoms, and I've just heard that the shrinking scientists are having a very tiny bit of bother…

LET'S TAKE A LOOK...

...DEEP INSIDE.

AWESOME ATOMS

Even though atoms are truly terribly tiny, they're still bigger than the start of the universe. In this crucial chapter we'll take a tiny peek inside an atom, and you can try cooking your own (all you need is a seriously small frying pan!).

The shrinking scientists … get a buzz from an atom

The shrinking scientists are inside an atom...

The **nucleus** is made of smaller bits of matter called protons and neutrons.

proton

Even more tiny blips of matter called **electrons** whizz around the nucleus.

THE ELECTROMAGNETIC FORCE PULLS THE ELECTRONS AND PROTONS TOWARDS EACH OTHER. IT STOPS THE ELECTRONS FLYING OFF.

VZZZ! VZZZZ!

neutron

BUZZ OFF!

HEY! MIND MY EARS!

BUT NOT ALWAYS...

Could you be a stunning scientist?

TRUE or FALSE? (Answers on page 13.)

1 If the nucleus was an elephant dropping, the electrons would be flies buzzing around it several metres away.

2 Your body is mostly empty space.

3 The electrical current that powers your CD player is made of neutrons.

4 If it wasn't for the electromagnetic force, you'd be a slimy puddle.

5 Thanks to the electromagnetic force, your bottom is floating above your chair.

IT'LL BE A **BIG JOB** TO SCOFF THAT LOT...

...YEAH, A MASSIVE **TUSK** – I MEAN TASK!

The complicated science bit

It's true, your bum really is floating! The electromagnetic force pulls electrons and protons together but it *pushes* electrons apart. Look what happens when two electrons whizz towards each other…

And that's why your bum is floating. Your backside and the chair are pushed apart by their electrons. What's more, it's the same when you sit on a toilet or try to touch anything – all you can feel is the force of the anti-social electrons shoving each other away. Mind you, the distances involved are so tiny it's easy to feel that you're making contact.

And now you know a bit about awesome atoms I bet you can't wait to make your very own! Turn the page for some crazy cookery, but be warned – you could blow up the planet!

Bet you never knew!
If atoms didn't push each other away, they could pass through each other – after all, they're mostly empty space. And that means you could walk through walls like a ghost!

Answers:

1 FALSE – The electrons would be several KILOMETRES away, and they'd be smaller than flies.

2 TRUE – Your whole body is made of atoms and each one is mostly empty space.

3 FALSE – It's made of electrons moving in an electric wire and it's *shockingly* dangerous – so don't mess with it!

4 TRUE – Without the fantastic force, your body's electrons would fly away and your atoms would fall to bits. So you wouldn't just be in the soup – you'd *be* the soup!

5 TRUE – And this page tells you the complicated science behind it…

HOW TO COOK YOUR OWN ATOMS

Are you itching to know how atoms formed? Well, you've come to the right place!
Atoms formed in the Big Bang and giant stars, and we're going to try cooking our own.

Quarks (rhymes with "larks") are stunningly small specks of matter that make up protons and neutrons. Quarks come in six varieties with interesting names.

 UP DOWN CHARM  STRANGE TOP BOTTOM

An **element** is a type of atom. There are 95 natural elements. Each element has a certain number of protons.

Horrible Science Cosmic Cook Book
Recipe 1: Hydrogen atom Big Bang-style

This simple little atom makes a tasty snack at any time of day —
if you can remember where you put it!

You will need
A mixing bowl, a terribly tiny frying pan, a spoon and a small, hot universe

Ingredients
1 up quark
2 down quarks
1 electron
Parsley
Salt and pepper to taste

WHAT'S UP?

WE'RE FEELING DOWN!

Method
1. Preheat your universe to 10 billion°C and set off a Big Bang.

2. Mix in the quarks and stir well.
3. Cook the quarks for a split second until the strong force glues them together to make a proton. Be careful — if you mix two up quarks and one down quark you'll be making a neutron by mistake!
4. Cool slowly for 380,000 years, add an electron and serve with the other ingredients added for taste.
CONGRATULATIONS — you've cooked a hydrogen atom! And that's how the first atoms formed!

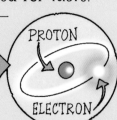
PROTON
ELECTRON

The Big Bang was a big success, but there was a problem. So far only three elements had formed – hydrogen, helium and a squitty squirt of lithium – but more elements were needed for life to begin. Luckily for us, about 200 million years later giant stars were blasting them out…

Horrible Science Cosmic Cook Book
Recipe 2: A feast of elements

Impress posh guests as you dish up a tasty atom treat! And what's more, it won't cost the Earth... Well, yes it will (see step 4).

You will need
A mixing bowl, a terribly tiny frying pan, a spoon and a giant exploding star

Ingredients
Hydrogen and tomato ketchup

Method
1. Switch on your star. Stars crush hydrogen protons to make helium. Some matter turns into heat and light energy (WARNING — this will keep your house nice and warm but it might burn down).
2. Use up the hydrogen and heat your star until it's 50 billion°C. This will take about 11 million years.
3. Relax as your star's gravity squishes the helium into new elements such as carbon, oxygen and nitrogen.
4. In 1,012,017 years the star will blow up and destroy the Earth. Oh well, you can invite your friends round for a big blow-out!

IS IT NEARLY READY? WE'RE STARVING!

JUST WAITING FOR THE STAR TO HEAT UP — IT'LL BE ABOUT 11 MILLION YEARS.

Get the idea? Most of your body's atoms were made in exploding stars billions of years before you were born. And that means you really are star quality!

I COME FROM OUTER SPACE?

I'M AN ALIEN!

EVIL ELEMENTS

Every element is different, but since this is a Horrible Science book we'll be checking out the evil elements that you *definitely* wouldn't want to find in your dinner… And where better to discover evil elements than the nasty notebook of a genuine crazy chemist?

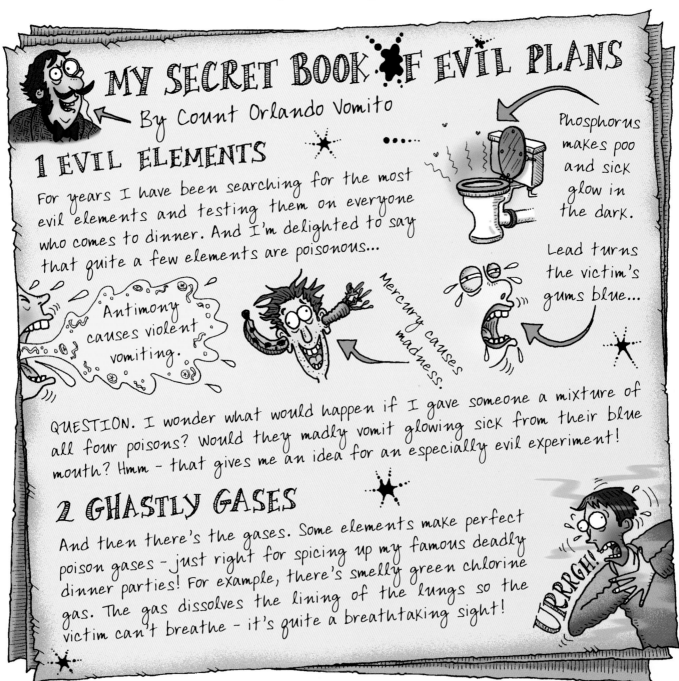

MY SECRET BOOK OF EVIL PLANS
By Count Orlando Vomito

1 EVIL ELEMENTS

For years I have been searching for the most evil elements and testing them on everyone who comes to dinner. And I'm delighted to say that quite a few elements are poisonous…

Phosphorus makes poo and sick glow in the dark.

Lead turns the victim's gums blue…

Antimony causes violent vomiting.

Mercury causes madness.

QUESTION. I wonder what would happen if I gave someone a mixture of all four poisons? Would they madly vomit glowing sick from their blue mouth? Hmm – that gives me an idea for an especially evil experiment!

2 GHASTLY GASES

And then there's the gases. Some elements make perfect poison gases – just right for spicing up my famous deadly dinner parties! For example, there's smelly green chlorine gas. The gas dissolves the lining of the lungs so the victim can't breathe – it's quite a breathtaking sight!

URRRGH!

3 ROTTEN RADIOACTIVITY

If I was feeling really evil, I could try making radioactive poisons. All I'd have to do is zap a nucleus of a radioactive element like uranium with neutrons. The zapped bits of uranium nucleus zap nearby uranium atoms, and the whole thing goes out of control and causes an atomic blast with lots of deadly X-rays and gamma rays! Now that would make my next party go with a bang!

YOU'RE INVITED TO MY RADIOACTIVITY PARTY
IT'LL COST ME A BOMB BUT YOU'LL HAVE A BLAST!

What's that? You think splitting the atom sounds like a cool experiment to try at home?! DON'T YOU DARE!!!!!!! Setting off an atom bomb in your bedroom is a very antisocial thing to do, as it's sure to flatten your town and give your hamster a nervous breakdown. And what's more, you'll have to clear up the mess and pay for the damage out of your pocket money.

Hmm – maybe it's better to wind up this weird and wacky topic with a few…

EXTRAORDINARY ELEMENT FACTS

1 The element tantalum is named after Tantalus. In Greek mythology, Tantalus was tortured in hell by being forced to stand up to his neck in water without being allowed to drink. The element was almost as hard to get hold of as a glug of water for thirsty Mr T.

TANTALUS HAVING TANTALIZING THOUGHTS

 2 The element francium is so rare that there could be just 20 francium atoms in the whole Earth. (There are 19 more francium atoms hiding in this book – so keep a look out for them!)

3 The most common element on Earth is oxygen. We need to breathe oxygen to stay alive. The second most common element, silicon, is completely useless for our bodies so don't go munching a silicon sandwich.

4 Most of your body is made of just three elements – 63 per cent of your atoms are hydrogen, 25.5 per cent are oxygen and 9.5 per cent are carbon.

Bet you never knew!
Your body needs a bit of potassium and sodium to help your nerves work properly, and you've also got a dash of sulphur that came from volcanoes.

But wherever atoms are found – at the end of your nose or at the end of the purple slobwobbler's left eyeball – they all have something in common. They make the electromagnetic force and that means they help you watch TV. Tune in to the next page to find out what the heck I'm talking about…

LETHAL LIGHT AND EXTREME ENERGY RAYS

The story so far… The electromagnetic force holds atoms together and makes your bum hover. These two pages are about the other things this force does for us.

A quick note from the author…
To make sense of the force, you need to read either the simple explanation or the horribly complicated scientific one involving varying electromagnetic wavelengths and blips of energy called photons. So – what's it to be?

THE SIMPLE ONE!

The simple explanation

Imagine the electromagnetic force as an energy ray. If you turn up the energy level you can make the ray into a TV signal, light or an X-ray, and lots more interesting things. So let's try a stunningly horrible experiment to show what this fantastic force can do…

THE EVIL EGYPTIAN MUMMY EXPERIMENT

OOOH!

GZZZZ!

LOW ENERGY = radio waves.

WELCOME TO THE HORRIBLE SCIENCE LAB! We're just about to fire our high-powered, laser-boosted electromagnetic ray blaster at this mummy…

HIGH ENERGY = UHF waves used for TV signals.

EH?

ERK!

EVEN HIGHER ENERGY = microwaves. Yes, the things you cook with.

MUCH HIGHER ENERGY = heat waves, or infrared as scientists call them. The mummy is still cooking but now he's crisping up nicely.

YOUCH!

OOER!

A BIT HIGHER ENERGY = ordinary everyday light. Yes, it's the stuff that shines from light bulbs. The mummy sees the ray as red light, but as we turn up the energy he'll see yellow, green and then blue light.

A BIT HIGHER ENERGY = ultraviolet light. The poor mummy gets a nasty sunburn from these high-energy rays.

YEOW!

HUH?

VERY HIGH ENERGY = X-rays. We can see the mummy's bones.

DEADLY DANGEROUS KILLER ENERGY = gamma rays.

YARGH!

Pfwoar – sniff that, readers?! The mummy got burnt. Or, as a scientist would say, some of the mummy's atoms bonded with oxygen atoms in the air to make new substances. It's chemistry, it's crazy, and there's a whole lot more of it happening in the next chapter…

CRAZY CHEMICAL CHAOS

Atoms are unfriendly objects. Most of the time they just bounce apart, but once in a while they stick close to one another and that's when you make a new substance.

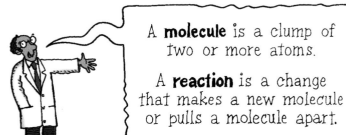

A **molecule** is a clump of two or more atoms.

A **reaction** is a change that makes a new molecule or pulls a molecule apart.

And thanks to this small bit of science-speak, the following facts should all make stunning sense – er, hopefully…

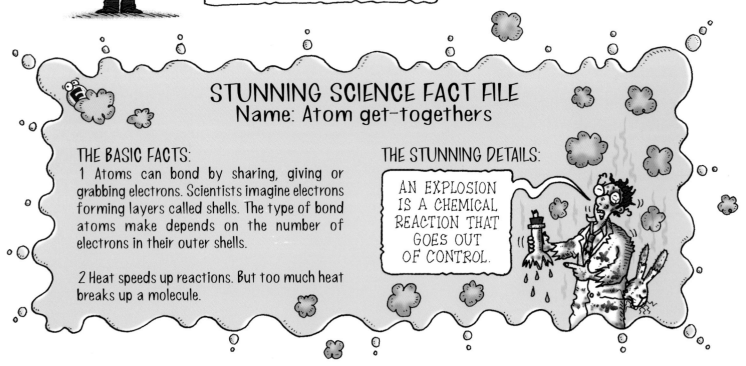

STUNNING SCIENCE FACT FILE
Name: Atom get-togethers

THE BASIC FACTS:

1 Atoms can bond by sharing, giving or grabbing electrons. Scientists imagine electrons forming layers called shells. The type of bond atoms make depends on the number of electrons in their outer shells.

2 Heat speeds up reactions. But too much heat breaks up a molecule.

THE STUNNING DETAILS:

AN EXPLOSION IS A CHEMICAL REACTION THAT GOES OUT OF CONTROL.

Getting in a state

To make things even more chaotic, atoms and molecules are always in a state… No silly, I don't mean they're tired and emotional. States of matter include a solid, a liquid, a gas or a plasma – it all depends on how *hot* the substance is…

Hmm – this sounds confusing! Tell you what, let's go back to the Horrible Science lab to see all this in action. The shrinking scientists are now back to their normal size and enjoying a well-earned cup of coffee whilst Mr Fluffy enjoys a well-earned snooze… Little do they know that we're about to use their coffee (presently in liquid form) to show you how heat makes a substance change state…

When we freeze the first coffee, the molecules form into a solid state called ice. This is impossible to drink…

When we heat the second coffee into a kind of gas called a vapour, the molecules float away. Of course you can't drink gas…

YEOW! MY TEETH!

CRUNCH!

I'M GETTING RATHER STEAMED UP!

We'll turn the third coffee into a plasma state. This is a super-heated gas, where electrons get torn away from atoms.

MAN – THAT'S EVEN WEIRDER THAN MY DREAM!

…DON'T FORGET TO DRINK YOURS!

WAAAH!

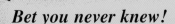

Bet you never knew!
1 In many solids, the atoms form in pretty shapes called crystals. In solid water, ice crystals form arty snowflake patterns and you can squish them down your sister's neck (in the interests of scientific research, naturally).
2 Diamonds are crystals made by carbon atoms. These crystals are incredibly hard, but scientists reckon that they lose their shape after billions of years. Hmm – trust a scientist to prove that diamonds AREN'T forever!

OH YUCK! What's that smell?! OH NO – it's the gasping gases leaking from the next page!

GASPING GASES

Welcome to two pages of ghastly gasping gas facts! We'll be sniffing out smelly substances and finding out why you're breathing the breath of dead people. But right now it's time for some more cosmic cookery…

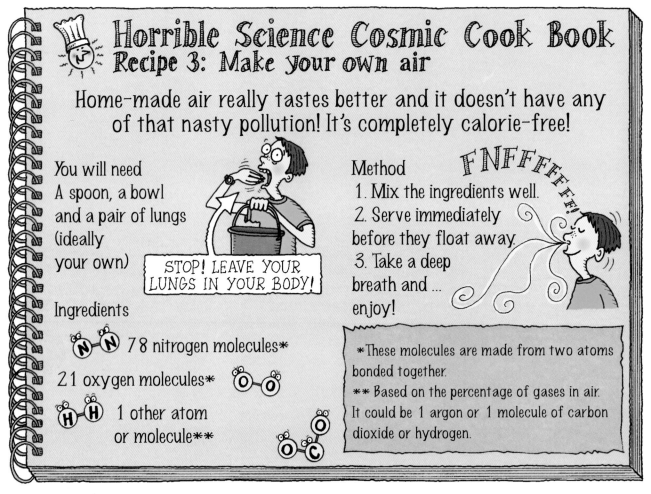

Horrible Science Cosmic Cook Book
Recipe 3: Make your own air

Home-made air really tastes better and it doesn't have any of that nasty pollution! It's completely calorie-free!

You will need
A spoon, a bowl and a pair of lungs (ideally your own)

STOP! LEAVE YOUR LUNGS IN YOUR BODY!

Ingredients

78 nitrogen molecules*

21 oxygen molecules*

1 other atom or molecule**

Method
1. Mix the ingredients well.
2. Serve immediately before they float away.
3. Take a deep breath and … enjoy!

FNFFFFFF!

*These molecules are made from two atoms bonded together.
** Based on the percentage of gases in air. It could be 1 argon or 1 molecule of carbon dioxide or hydrogen.

Let's take a really, really close look at a container of air through our very expensive mega-microscope…

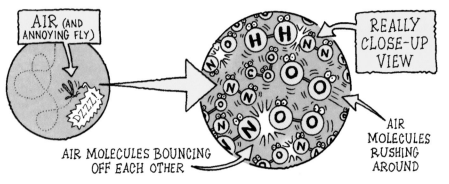

AIR (AND ANNOYING FLY)

REALLY CLOSE-UP VIEW

AIR MOLECULES BOUNCING OFF EACH OTHER

AIR MOLECULES RUSHING AROUND

It looks completely crazy! But you'll probably gasp when I say the inside of your lungs looks like this every time you take a breath of air. And now read on for another gasp-worthy fact...

And now for a gas molecule that you definitely wouldn't want to breathe...

Messy methane

Methane is a molecule made from one carbon and four hydrogen atoms. That's why chemists call it CH_4. The molecule is actually made by microbes as they rot plants in swamps and in the guts of animals and humans. And that's why methane turns up in farts.

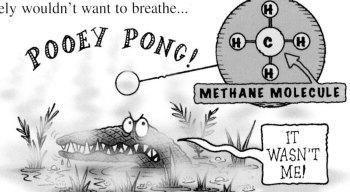

POOEY PONG!

METHANE MOLECULE

IT WASN'T ME!

METHANE – THE GOOD, THE BAD AND THE UGLY

The good: Your gas cooker probably burns methane. In fact, if you had a dinosaur-sized bottom problem you could convert your toilet into a gas cooker.

The bad: Cows produce more methane than we do. Scientists reckon a cow farts almost 200 grams of the gas every day and the world's cows parp over one hundred million tonnes of methane every year.

PARP!

DO YOU WANT BROWN SAUCE WITH YOUR SAUSAGE?

SIZZLE!

ER – NO THANKS!

CH_4! AWESOME!

The ugly: That's ugly news for the world. Methane and other gases such as carbon dioxide trap heat in the Earth's atmosphere (the air surrounding the planet). As the Earth gets hotter the climate may change in violent ways, with lots of storms and floods and droughts and horribly hot weather. Scientists call these changes the greenhouse effect – but it's nothing to do with the glass shed at the end of your garden.

WEIRD WATER AND LOOPY LIQUIDS

We've been wild and windy, and now it's time to get runny with liquids. In fact any element is a liquid at the right temperature – and to prove it, here are some liquids you wouldn't want to paddle in…

Weird water wonders

But of course, the liquid everyone thinks of is water, so we've asked the shrinking scientists to take a really close look at the water molecules in this toilet…

The shrinking scientists in … toilet trouble

LET'S TAKE A CLOSER LOOK AT THIS WATER MOLECULE...

HYDROGEN

OXYGEN

HYDROGEN

THE WATER MOLECULE IS MADE OF ONE OXYGEN AND TWO HYDROGEN ATOMS – THAT'S WHY SCIENTISTS CALL IT H_2O.

WEIRD WATER FACTS

1 The pressure of the water pressing on your body is called (howls of amazement here) water pressure. The force gets stronger the deeper you go. Traditional diving suits had air pumped into them to withstand horribly huge water pressure – but if the air pumps failed, the force was said to squeeze the diver's body up the air tubes like toothpaste. Only the bones were left in the suit.

SHLOOP!

2 The hydrogen protons in a water molecule pull on neighbouring molecules and make them form drippy drops or a springy surface that bugs practise ballroom dancing on.

3 But water isn't so friendly to other molecules. The hydrogen protons pull other molecules to bits until they dissolve and that's what happens to the sugar in your tea...

Your body is a wobbly bag full of water (roughly two-thirds of your body is water and your brain is 80 per cent water – which means it's as watery as a potato). Not surprisingly, you'll drink about 73,000 litres of water in your lifetime – enough to fill 292 baths. And if you didn't drink anything for a couple of weeks, you could die in a stunningly foul fashion.

In 1905 Pablo Valencia was searching for gold in the Arizona desert. But he got lost, and for seven days the only water he drank was his own wee. Here's what seven days of terribly torturing thirst did to poor Pablo's body...

Mind you, if there's one thing worse than not drinking any water – it's drinking too much! If you happened to drink a whole bath of water in an hour, you'd disrupt your nerve signals and probably die. Wat-er choice! You can dry out and die out, or gulp a few buckets and kick the bucket...

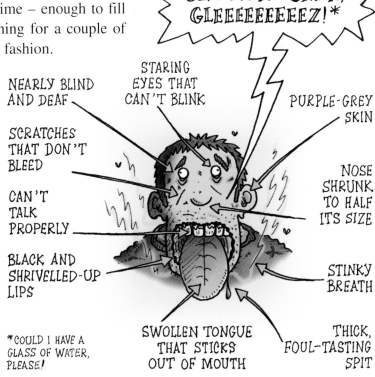

GURD-EYE AV-HAY GLA-ZOFF ORKA, GLEEEEEEEEZ!*

NEARLY BLIND AND DEAF

STARING EYES THAT CAN'T BLINK

PURPLE-GREY SKIN

SCRATCHES THAT DON'T BLEED

CAN'T TALK PROPERLY

NOSE SHRUNK TO HALF ITS SIZE

BLACK AND SHRIVELLED-UP LIPS

STINKY BREATH

*COULD I HAVE A GLASS OF WATER, PLEASE!

SWOLLEN TONGUE THAT STICKS OUT OF MOUTH

THICK, FOUL-TASTING SPIT

THE WONDER OF YOU

Awesome, isn't it? Somehow a collection of molecules got together and arranged themselves to make you. And it's stunning to think that you're only alive because such tiny things do their jobs. "But what are they up to?" I hear you ask. Well, as luck would have it, you've come to the right page to find out…

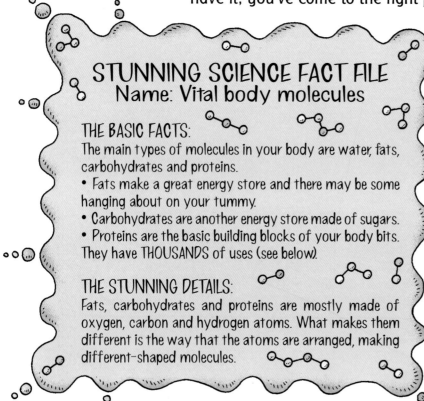

STUNNING SCIENCE FACT FILE
Name: Vital body molecules

THE BASIC FACTS:
The main types of molecules in your body are water, fats, carbohydrates and proteins.
• Fats make a great energy store and there may be some hanging about on your tummy.
• Carbohydrates are another energy store made of sugars.
• Proteins are the basic building blocks of your body bits. They have THOUSANDS of uses (see below).

THE STUNNING DETAILS:
Fats, carbohydrates and proteins are mostly made of oxygen, carbon and hydrogen atoms. What makes them different is the way that the atoms are arranged, making different-shaped molecules.

Bet you never knew!
1 If you chew bread long enough it tastes sweet. That's because bread contains carbohydrates and when you chew, your spit breaks them down into the sugars that make them up.
2 Tough insect shells are also made of carbohydrates. Any volunteers fancy chewing a nice crunchy dead cockroach?

And now let's take another peek at proteins. These marvellous molecules are so great, we've given them free advertising space…

PROTEINS – THEY'RE THE ONES FOR YOU!

• Feeling more dead than alive?
• Could you use a bit of help?
• The protein squad are here for YOU!
• Thousands of complicated molecules for thousands of important jobs! Enough to make up three-quarters of all the solid material in your body!

HI!

Order now and get free collagen to build your skin and bones!!!
PLUS free fingernails and toenails
(they're made from the protein keratin)!!!

I GUESS EVERY**BODY** NEEDS THEM!

DIGESTIVE PROBLEMS?

Don't suffer in silence! Use our protein enzymes to digest your food! The unique shape of an enzyme provides a snug site for chemical reactions ... and it can do it 1,000 TIMES A SECOND!

STEP 1: Food molecule fits into enzyme...

STEP 2: Ta-da! It's broken in two!

THANKS!

Eny zyme!

THE SMALL PRINT Enzymes are destroyed by heat and acid, so don't take a bath in boiling acid!

A bit of good news…

You'll be relieved to read that you can buy proteins from any supermarket – yes, they're found in foods such as meat, beans, cheese and fish. And all the time you're reading this book, your busy body is breaking down old proteins and making new ones from smaller molecules called amino acids.

Brainy Boffin

Antoine-François Fourcroy (1755–1809)

One of the first scientists to study the body's chemistry was Frenchman Antoine-François Fourcroy. Fearless Fourcroy looked at the effects of heat, air, water, and also acids and other chemicals on rotting human corpses. Oh well, at least he found a useful body of evidence.

Bet you never knew!

1 When a body rots, atoms escape from its proteins. The atoms are taken up by plants and eaten by animals, and eventually turn up in your dinner. So that means that your body contains atoms from mouldy dead bodies.

2 There's a kind of stinky meteor from space which actually contains amino acids. So were they sent by aliens?

Oh dear – the shrinking scientists aren't too happy with the idea of aliens…

YOU CAN'T SAY THAT!

THERE'S NO PROOF ALIENS EXIST!

THE CHEMICALS FORM NATURALLY IN SPACE.

Oh, all right! All we can safely scientifically say is that billions of years ago, amino acids arrived on Earth and somehow formed into murderous microbes. And there's loads of them lurking in the next chapter…

MIGHTY MICROBES

Slowly but surely the terrible topics in this book are getting bigger. I mean, microbes are bigger than molecules, even if they are too small to see without a microscope. But maybe that's not such a bad thing, because right now they're wriggling up your nose and other places I wouldn't like to mention…

Microbes live all over the world and can be found virtually anywhere – in the air, earth, water, plants, animals, and even on you. Here's what you're looking for…

MIGHTY MICROBE-SPOTTER'S GUIDE

Name/description of microbe	Stunningly small size	Loathsome lifestyle
VIRUSES DNA skulking inside a protein shell.	A line of 200,000 would stretch 1 cm. But that's hundreds of times bigger than atoms.	They sneak into the body cells and force them to make new viruses. They cause diseases such as colds – atishoo!
BACTERIA Slimy thingies without a special place to keep their DNA.	About 50,000 to 1 cm. The biggest bacteria are 1 mm long!	The biggest bacteria are blue-green and spend their lives in the ocean scoffing sulphur.
PROTISTS (some are known as protozoa).	100 to 1 cm.  THAT'S MASSIVE!	Some guzzle bacteria or other protists. Others make food from light and chemicals, in the same way as plants (see page 46).

Cells aren't always rooms to lock bad people in! Cells are tiny living units that make up most living things — including you! (For more details, take a close-up look at pages 46 and 57.)

DNA is the chemical code for building life forms (see page 58). DNA is stored in cells.

Right now the shrinking scientists are checking out the microbes in one of their most horrible hideaways … the mouth of New York private eye MI Gutzache…

The shrinking scientists in … the microbe mouth mission

Bet you never knew!

1 Stop making faces, readers! There are probably protists in your mouth too – and on your toothbrush. This microbe happily lives in dogs' mouths too – I wonder if Gutzache got his protists from a slobbery doggy kiss?

2 You can combine the cell wall, nucleus and jelly-like body of three different protists to make a new one. Just imagine a mad scientist using the skin, brain and body bits of three people to make a new human…

WHO ARE YOU LOOKING AT?

Meanwhile, back in MI Gutzache's mouth…

Every protist is a marvel of micro-engineering, with over 50,000 different proteins and more parts than a jumbo jet. But the most amazing bits could well be the mitochondria. These thingies are mini power stations making energy from food chemicals – and guess what? You've got them too! It's stunning to think you've got things in common with a slimy protist…

WELCOME TO THE SLIME ZONE!

The shrinking scientists are back in the Horrible Science lab to test our new Horrible Science bacteria breeder…

Bacteria breed by splitting in half – doubling their numbers every 20 minutes. Oh well, let's leave them doubling at the double while we take a look at this microbe newspaper…

In this week's issue
▷ Bacteria beauty tips
▷ Gourmet garbage guide
▷ Staying healthy by giving humans diseases

THE DAILY MICROBE
The world's smallest paper!

BACTERIA IN DEODORANT ARMPIT TERROR!
by rotten reporter Sly M Ball

This morning 2.6 million bacteria were put in danger when a human dabbed antiperspirant on his armpits. There we were enjoying a lovely drink of sweat, with all its healthy vitamins and minerals, when we got splatted. Thousands of bacteria died in the terror. I escaped by clinging to a giant rope (well, actually I think it was a hair).

The editor writes...
What is it with humans? These days they keep trying to spray and scald and bleach us out of existence! Anyone would think they don't like us!

OH NO! While we've been reading the paper, the bacteria have been breeding out of control!!! Yes, in just one day a bacterium can breed 280,000 BILLION more bacteria!

Now there's only one thing to do… PANIC! And after we've panicked, let's call in the Bacteria-Buster Company.

HOW TO GET RID OF BACTERIA IN SEVEN EASY STEPS (By the BBC)

1 Put this book in the freezer. Some bacteria will freeze to death. BRRR!

GLUG! **2** Give the book a soapy bath. Soap and water wash bacteria down the plughole, and they're great for cleaning skin.

3 Leave the book to dry in the sun. Ultraviolet rays kill some bacteria. EEK!

4 Poison the bacteria — chlorine or bleach dissolves their little bodies.

HORRIBLE HEALTH WARNING
It also dissolves your little body so, on second thoughts, don't mess with this stuff.

5 Chop the book into tiny bits smaller than bacteria, and pass the bits through a fine filter to strain out bacteria. YOUCH!

6 Blast the book with a high-pitched sound. The bacteria will wobble until they explode.

SOUNDS SHOULD BE HIGHER THAN THIS.

7 Zap the book with gamma rays (beware — bacteria can take 750 times more gamma rays than you can).

Well, that should do the trick… You can read on now!

UGH!

EESH! I'M OUTA HERE!

PULL THE CHAIN!

HARD!

CHEERS!

DEADLY DISEASE DETAILS

Even though microbes are horribly tiny, they have the power to spoil your whole day – and I'm not talking about smelly armpits or bad breath. As you're about to find out, some microbes cause deadly diseases… Er, let's hope you don't catch anything too nasty from these pages!

Here's how microbes can be murderous…
• Bacteria can cause food poisoning, blood poisoning, plague and acne.
• Protists can cause typhoid, dysentery and malaria.
• Viruses cause influenza, colds, chicken pox, rabies, mumps, yellow fever and a host of other horrors…

CONGRATULATIONS, MR JONES – YOU'VE GOT EVERYTHING!

GROAN!

It's a terrible thought that something the size of a human can be destroyed by something as tiny as a microbe. In 1922 one of the nastiest microbes ever struck a Scottish hotel – but which murdering microscopic monster did the deadly deed?

The notebook of Inspector McKitchen

GASP! GASP! GASP!

DRIBBLE!

THE CASE OF THE HOTEL OF HORROR

I was called in when guests at the hotel in Loch Maree began dropping dead. Each victim developed double vision. They found it hard to move and breathe, and dribbled non-stop. Death followed in hours. At first I suspected poison…

The victims had been out on a fishing trip the day before, so I looked at their lunch menu…

On questioning the other guests I found out that all the victims had eaten the duck-paste sandwiches. I suspected someone had dosed the duck – it looked to me like fowl play. But who? The kitchen staff swore that the jar had been sealed before the sandwiches were made. None of the staff had any reason to bump off the guests.

Was it poison at all, I wondered.

What if some kind of germ was involved?

The fiendish killer puzzle was giving me a horrible hammering headache. What could the answer be?

MENU FOR PACKED LUNCH

Choice of sandwiches…

Chicken and ham paste

Beef

Wild duck-paste

Bottle of beer

THROB!

McKitchen may be a figment of my imagination but the rest of the story is true. Six guests and two boatmen died – but what kind of microbe had done them to death?

a) A virus spread by sneezing.

b) A poison made by bacteria.

c) A poisoning protist that lived in the loch.

Answer:

b) A virus would spread to everyone in the hotel. A protist in the loch would have attacked everyone on the boat trip. So that leaves only the suspect in the WANTED FOR MURDER poster!

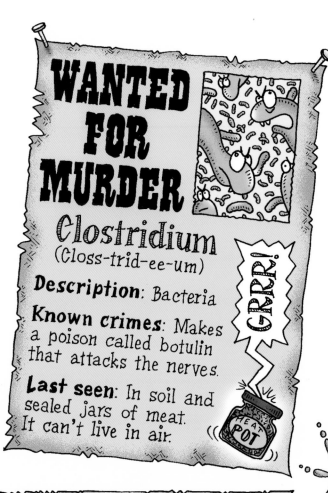

WANTED FOR MURDER

Clostridium
(Closs-trid-ee-um)

Description: Bacteria

Known crimes: Makes a poison called botulin that attacks the nerves.

Last seen: In soil and sealed jars of meat. It can't live in air.

GRRR!

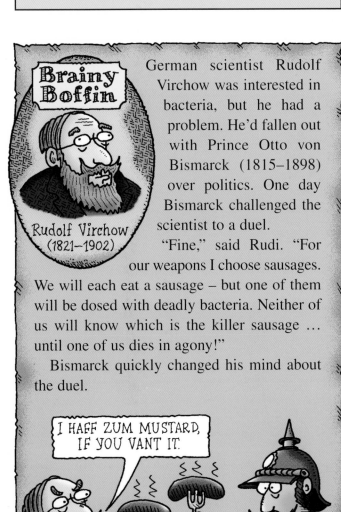

Brainy Boffin

Rudolf Virchow (1821–1902)

German scientist Rudolf Virchow was interested in bacteria, but he had a problem. He'd fallen out with Prince Otto von Bismarck (1815–1898) over politics. One day Bismarck challenged the scientist to a duel.

"Fine," said Rudi. "For our weapons I choose sausages. We will each eat a sausage – but one of them will be dosed with deadly bacteria. Neither of us will know which is the killer sausage … until one of us dies in agony!"

Bismarck quickly changed his mind about the duel.

I HAFF ZUM MUSTARD, IF YOU VANT IT.

Bet you never knew!
Botulin poisoning is incredibly rare. And that's good news because botulin is so deadly that a poison pill weighing 500 times less than this page could kill ***EVERYONE ON EARTH!***

DUCK-PASTE SANDWICH, INSPECTOR?

MAYBE NOT.

AT HOME WITH THE MITES

Before we escape from the wicked world of very tiny creatures, there are a few little friends I'd love you to meet. Say hello to the dust mite family… Aren't they lovely?

The half a house of horror

We've cut this nice little family home in half with a huge chainsaw to show your mite mates' hideaways…

QUICK QUEASY QUIZ QUESTION

So how exactly do you get rid of mites?
a) You can't.
b) Use a special micro-explosive called DIE-NOW-MITE.
c) Hoover them up.

Answer:
a) Mites are mite-y hard to get rid of! Even if you hoover a carpet for 20 minutes you won't suck up many mites, and the ones you catch live happily in the hoover bag. Washing your pillow gets rid of mites but more take their place because your house is jumping with them.

DON'T PANIC! You've been sleeping with mites all your life and they've never done any harm (probably). So sleep tight and be glad the mites don't bite. Not like the brutal bloodthirsty bugs in the next chapter…

Glad to hear it! You see, things are going to get awfully ugly. We'll be meeting some of the ugliest bugs ever, and we'll be hearing the worst about their even uglier habits...

Bloodthirsty bug families

So without further ado, let's meet a few bug families. But we're not talking happy families here. Some bugs enjoy eating other family members – hope your family isn't like this!

A **family** is a group of similar species (see page 45 for the full family facts).

A **species** (spee-shees) is a type of living thing such as an animal.

A grovelling apology from the author... Sorry, readers, owing to a printing problem, we've lost some words from these facts about bugs. Can you work out where the missing words fit in? Oh yes, and one missing word doesn't fit in at all! Confused? You will be!

Missing words:

Cockroach   Head

Screech   Tooth

Eyeball   Poo

Fangs   Snot

1 Wicked worms

Earthworms form the favourite food of munching moles. A mole bites off the worm's _____ and stores the rest of its body in an underground "pantry".

2 Munching molluscs

Slugs and snails slither along a carpet of slime called mucus (mew-cus). Imagine skating on slippery _____. The mucus tastes disgusting but hungry hedgehogs often scoff slugs and snails for supper.

3 Weird woodlice

Woodlice are related to crabs and lobsters, but you probably wouldn't want one in your seafood salad. They come out at night and feed on rotting wood and plants. Bacteria in their guts help them digest their food but they eat their own _____ to get the full goodness.

4 Mysterious millipedes

Here's a millipede love tip. If you want a girlfriend, why not bang your head on the ground or let out a loud _____. You can hear the musical millipedes 5 metres away. Anyone for singing lessons?

5 Cruel centipedes

IT WAS HORRIBLE!

Centipedes love chasing millipedes and killing them with their fearsome fatal poison fangs. In 1972 a centipede was found in a woman's _____. I bet she felt down in the mouth.

6 Scary spiders

There are 35,000 spider species, but they've all got eight legs, poisonous _____ and an urge to share your bath. Oh yes, and they all eat other creatures. Message to any flies reading this – YOU HAVE BEEN WARNED!

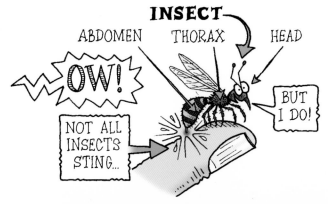

INSECT

ABDOMEN   THORAX   HEAD

OW!

NOT ALL INSECTS STING...

BUT I DO!

7 Irritating insects

All insects have six legs and three body bits…
But they come in a stunning selection of species, including ants, bees, wasps, butterflies, moths, beetles, and not forgetting the cute and cuddly _____.
In fact there's enough to keep a bug-brained boffin busy for years. (Call them entomologists (en-toe-mol-lo-jists) if you want to sound seriously scientific.)

Bet you never knew!

Bugs can be big…
• Goliath beetles can be 12 cm long and weigh 100 grams. African children tie string to them and fly them like model planes.
• Some stick insects grow to 33 cm long.
• The Queen Alexandra butterfly is 28 cm from wing tip to wing tip.

And bugs can be really tiny…
There's a type of patu spider in Western Samoa that's only the size of a full-stop. Try being scared of that!

GRRR!

THE SHRINKING SCIENTISTS

EEK! A PATU SPIDER!

UGH! IT'S MASSIVE!

HELP! IT'S GONNA EAT US!

Answers:
1 Head. But if the mole bites its tail, the worm can grow a new one and escape! 2 Snot 3 Poo 4 Screech 5 Tooth 6 Fangs 7 Cockroach. The word that you didn't need is eyeball.

THE GRUESOME GARDEN GUIDE

AMAZON ANTS raid other ants' nests and steal their eggs. The babies are brought up as slaves of the awful Amazon ants.

Beware of **AUSTRALIAN BULLDOG ANTS!** Thirty bites can kill a human – and what's more you get a lovely free squirt of acid in your wounds.

LEAFCUTTER ANTS are keen on gardening. They snip up leaves with their jaws and grow mouth-watering mould on them. And to make their gardens grow better they add a dollop of fertilizer – their own poo.

The **ROSETHORN TREEHOPPER** hides from hungry bugs by pretending to be a ... well, any guesses? For extra accuracy it points its body in the direction of the real thorns. The bug sucks juices from the roses – so I guess it's a thorn in the rose's side.

The **TORTOISE BEETLE** looks like ... NO! Not a tortoise! But its odd shape helps it to hide amongst fallen leaves.

The **KING SWALLOWTAIL BUTTERFLY CATERPILLAR** looks just like ... NO! Not a king – it looks like a bird dropping. A bird thinks, "It may be a juicy caterpillar or it may not. Better not risk it!"

But talking about poo – the **ASSASSIN BUG** eats termites and sneaks up on them disguised as a pile of termite poo. It sticks the poo on to its back for this job. Now I don't know about you, but if I was being chased by a giant pile of poo I think I might just notice.

The **RAILROAD WORM** is a South American beetle grub that scares other bugs away by making lights on its sides and even a red light on its head. It looks like a little train, although you wouldn't go far if you sat on one.

DUNG BEETLES bury poo and lay their eggs in it. This is a very handy service for larger animals because it keeps the grass nice and clean and beetles are happy because their yucky young dine on delicious dung. One species of dung beetle can even sniff out buffalo dung when it's still in mid-air. And they're on their way before it hits the ground...

HUMAN BOT FLIES lay their eggs on mosquitoes, but when the mosquito bites a human the gruesome bot-fly grubs burrow into our skin. If a grub's at home on your head you can hear it feasting on your flesh.

LOATHSOME INSECT LIFESTYLES

By now you might be thinking that insects are annoyingly antisocial – but that's not the half of it. Just look at their loathsome lifestyles!

A TYPICAL INSECT'S LIFE STORY

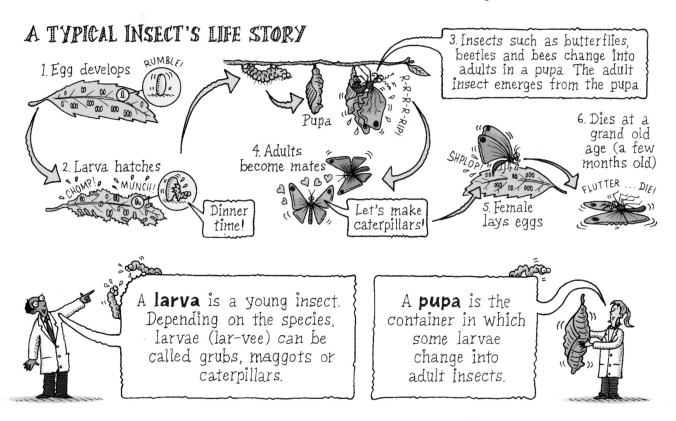

1. Egg develops

RUMBLE!

2. Larva hatches

CHOMP! MUNCH!

Dinner time!

Pupa

4. Adults become mates

Let's make caterpillars!

R-R-R-R-RIP!

3. Insects such as butterflies, beetles and bees change into adults in a pupa. The adult insect emerges from the pupa.

6. Dies at a grand old age (a few months old)

SHPLOP!

5. Female lays eggs

FLUTTER ... DIE!

A **larva** is a young insect. Depending on the species, larvae (lar-vee) can be called grubs, maggots or caterpillars.

A **pupa** is the container in which some larvae change into adult insects.

I could tell you loads of foul facts about every stage of an insect's life, but sadly I'd need a bigger book to fit them all in. Instead let's look at the mating stage. Mating spells deadly danger for male bugs and I'm sure insect magazines are full of warnings…

LOVE TIPS FOR UGLY BUGS
By Aunty Ant

Well, my dears, a bit of breeding is always a good idea, but do be careful. If you're a male praying mantis you must NEVER upset the female when mating. If she gets cross she might bite your head off. No, I don't mean shout at you – I mean she'll really bite your head off! And eat the rest of you afterwards!

So, my dears, you should try to be romantic, like the male dance fly, and offer your lady love a juicy dead insect wrapped up in silk you made yourself. But be warned – if she likes it she'll eat it, but if she doesn't she'll probably eat YOU! Mind you, if she's really hungry she'll scoff you as dessert anyway!

COME HERE, MY LITTLE PUDDING!

Horrible antisocial bugs

OK, so bugs eat other bugs, but do we care? I don't think so! But it's quite another story when bugs harm humans. It's bad enough when they guzzle our crops, but when they suck our blood it's completely out of order! Let's check out some bloodthirsty villains…

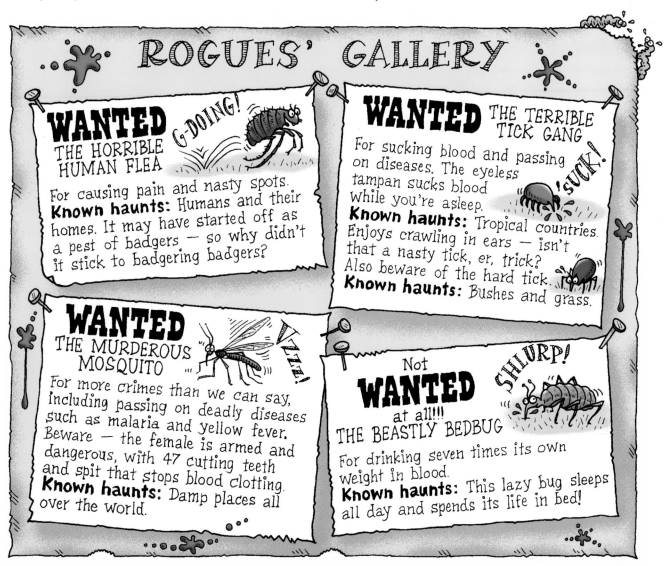

ROGUES' GALLERY

WANTED *G-DOING!*
THE HORRIBLE HUMAN FLEA

For causing pain and nasty spots.
Known haunts: Humans and their homes. It may have started off as a pest of badgers — so why didn't it stick to badgering badgers?

WANTED THE TERRIBLE TICK GANG

For sucking blood and passing on diseases. The eyeless tampan sucks blood while you're asleep. *'SUCK!'*
Known haunts: Tropical countries. Enjoys crawling in ears — isn't that a nasty tick, er, trick? Also beware of the hard tick.
Known haunts: Bushes and grass.

WANTED
THE MURDEROUS MOSQUITO

For more crimes than we can say, including passing on deadly diseases such as malaria and yellow fever. Beware — the female is armed and dangerous, with 47 cutting teeth and spit that stops blood clotting.
Known haunts: Damp places all over the world.

Not **WANTED** *SHLURP!*
at all!!!
THE BEASTLY BEDBUG

For drinking seven times its own weight in blood.
Known haunts: This lazy bug sleeps all day and spends its life in bed!

Bet you never knew!
In the 1970s US scientists tried, and failed, to put tiny radio transmitters on bedbugs and send them to spy on enemy forces in Vietnam. Bet you can't say "brainy boffins' biting bedbug bugging brainwave backfired badly" three times with a mouthful of bugs.

THEY MIGHT SEEM TINY TO YOU!

So does the insect lifestyle appeal to you? Thought not – but even if you don't like the way bugs live, you've got to admire their amazing abilities…

THE UGLY BUG OLYMPICS

It's a fact that bugs can do fantastic feats - as you're about to find out!

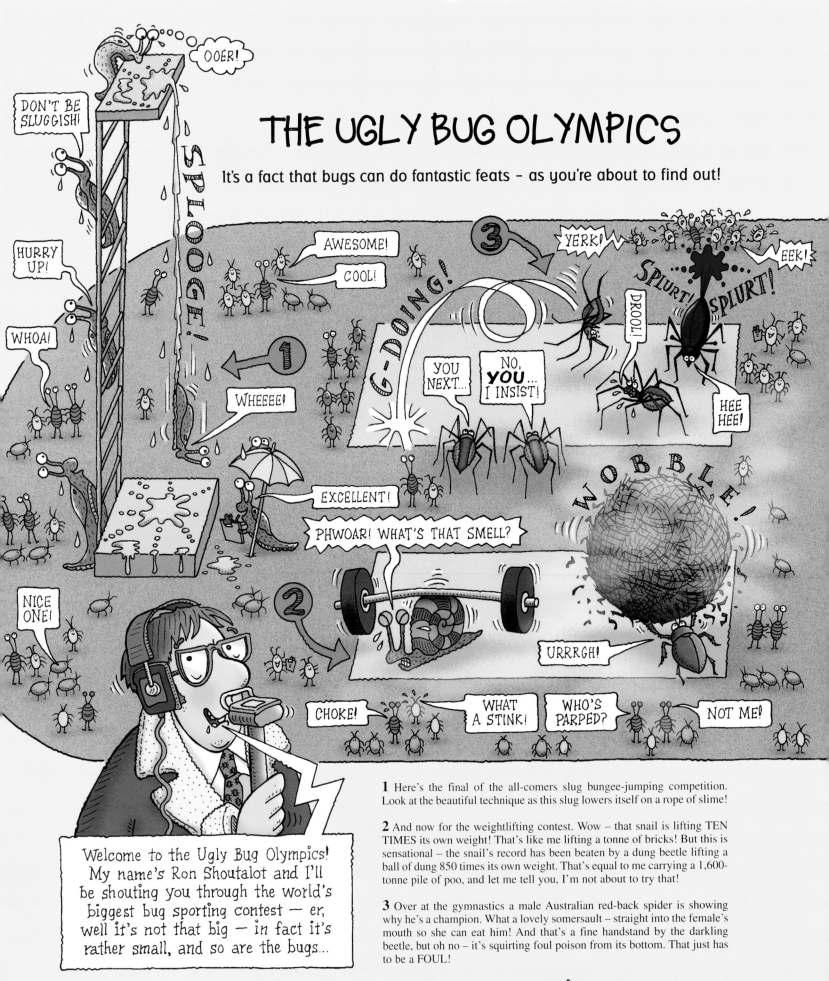

1 Here's the final of the all-comers slug bungee-jumping competition. Look at the beautiful technique as this slug lowers itself on a rope of slime!

2 And now for the weightlifting contest. Wow – that snail is lifting TEN TIMES its own weight! That's like me lifting a tonne of bricks! But this is sensational – the snail's record has been beaten by a dung beetle lifting a ball of dung 850 times its own weight. That's equal to me carrying a 1,600-tonne pile of poo, and let me tell you, I'm not about to try that!

3 Over at the gymnastics a male Australian red-back spider is showing why he's a champion. What a lovely somersault – straight into the female's mouth so she can eat him! And that's a fine handstand by the darkling beetle, but oh no – it's squirting foul poison from its bottom. That just has to be a FOUL!

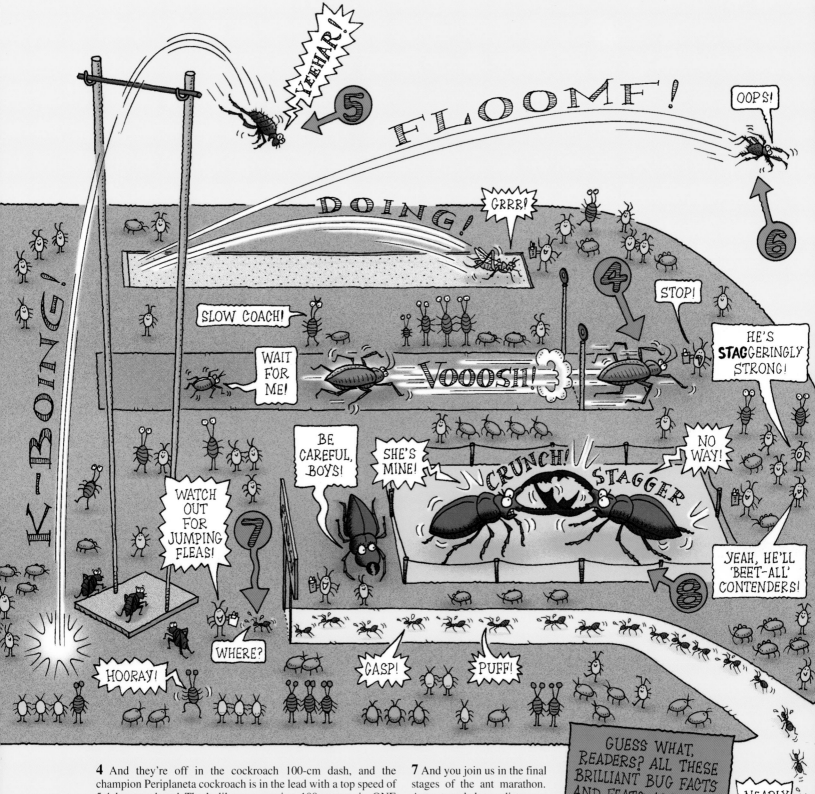

4 And they're off in the cockroach 100-cm dash, and the champion Periplaneta cockroach is in the lead with a top speed of 5.4 km per hour! That's like me running 100 metres in ONE SECOND! I think I'll pass on that one!

5 Great news from the all-comers high jump, where a flea has set a new world record of 130 times its own height. That's like me leaping five times higher than the leaning tower of Pisa! And that's not too likely!

6 I CAN'T BELIEVE MY EYES! There's drama at the long jump, where a grasshopper's personal best of 20 times its own body length has just been beaten by a jumping spider with – get this – FORTY TIMES its own length. That's like me jumping over THREE TENNIS COURTS!

7 And you join us in the final stages of the ant marathon. Ants travel long distances (compared with their size) in search of food, and these ants are reaching the end of a 3-km course. For me that's like a 60-km marathon! Phew – just thinking about it is bad enough!

8 Meanwhile, the beetle wrestling is hotting up as two male stag beetles lock jaws as they try to throw each other off a twig to win a female. Things could turn ugly if the loser lands upside down and gets eaten by the ants!

Got your breath back? Great – let's get to grips with some bigger beasts. The next chapter's really WILD!

AWFUL ANIMALS
AND PESKY PLANTS

Welcome to the nasty dog-eat-dog and cat-eat-mouse world of nature, in which plants and animals make a living by making other creatures do the dying. But first let's see how scientists sort out all the millions of life forms on planet Earth...

The BIG sort-out

You can imagine the scientific sorting system for living things as a set of Russian dolls – an animal or plant fits into a larger group, which fits into a larger group, and so on...

But who on Earth was brainy enough to think it up in the first place?

Brainy Boffin

Carl Linnaeus
(1707–1778)

The whole scientific sorting system was invented by Swedish scientist Carl Linnaeus – but careless Carl made mistakes. He believed tall tales about weird and wonderful creatures, and even made space in his system for humans with tails and humans who crawled around on all fours. Know any humans like that?

I'VE SEEN TWO LIKE THAT, MISTER!

EXCELLENT! I'LL STICK HIM IN WITH HORSES AND PONIES.

Bet you never knew!
Even after 250 years of scientific sorting, scientists reckon that 97 per cent of all living things are unknown to science! Take fungi, for example (that's moulds and mushrooms to you). Scientists think there are at least 70,000 different species, but there could be 1.8 million ... or more!

Getting it sorted
Anyway, the shrinking scientists are about to show how scientists classify a life form. Can we have a volunteer?

OK, so it sounds a bit of a performance, but there are millions of species in all shapes and sizes, and without some sort of sorting system, scientists would be in a horrible state of confusion. Talking about sizes, let's finish this page with some little and large animals.

Bet you never knew!

1 In 2004 US scientists discovered the stout infantfish of the Great Barrier Reef off the coast of Australia. It's just 7 mm long.

2 The blue whale is 4,000 times bigger. It's the biggest animal that ever lived – and that includes the dinosaurs. The 130-tonne whale weighs as much as 2,000 people. Its heart is as heavy as a car and even its tongue weighs as much as an elephant. If your tongue was that heavy, I bet you'd be tongue-tied…

And on that note, let's sneak back to the gruesome garden…

RETURN TO THE GRUESOME GARDEN

Sadly most of the bugs have escaped from our gruesome garden, but at least the plants are still here... OH NO – who let Mr Fluffy inside?! Oh well, here's our chance to compare him with a lettuce...

SPOT THE DIFFERENCES BETWEEN PLANTS AND ANIMALS

Plants don't walk around. They're rooted to the spot and their roots collect water from the soil.

Animals can't stay still.

Plant cell walls are made from cellulose.

Animal cell walls are made from a type of fat.

Plants make food in green leaves by photosynthesis.

Animals munch plants.

Plants give out oxygen gas.

Animal waste is poo and wee.

Cellulose (sell-you-loze) is a tough carbohydrate found in plants.

Photosynthesis (foe-toe-sinth-uh-sis) is the way plants make food using sunlight, water and carbon dioxide from the air.

Bet you never knew!
Plants come in all shapes and sizes. Wolffia – a type of duckweed, can be 0.5 cm across but Californian redwood trees grow to over 112 metres.

Anyway, let's see what's growing in our garden... Ooer – Mr F had better beware! It's full of vicious vegetables and painfully poisonous plants!

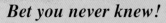

1. The **baobab seedling** can only grow if the thick coat of the seed is partly worn away. And for this to happen, it must pass through the guts of a baboon and come out in a dollop of poo.

2. This pretty **tamarisk** plant is collecting salt in its leaves. It drops its leaves to poison the soil so other plants can't grow nearby. With neighbours like that, who needs enemies?

3. The **South American strangler fig** strangles trees and steals water from their roots. Its victims die very slowly.

4. The **dead horse arum**. This foul flower smells like a dead horse. Flies get trapped inside, where many die. The surviving flies get dusted with pollen and take it to another awful arum. By the way, pollen contains plant DNA and it's needed to make seeds and fruits.

5. Do you drink coffee or tea? Coffee and tea plants make caffeine – the chemical that gives you a buzz. **South American holly** contains so much caffeine that the traditional way to drink holly tea is to vomit it up before it buzzes your brain too much.

6. When a foolish fly (or shrinking scientist) touches the trigger hairs on a **Venus flytrap**'s leaf, liquid rushes out of special cells. The leaf folds shut and produces acid. The trapped fly is slowly turned into slime and eaten alive.

7. Sticky **sundew** leaves curl around to trap and eat insects. But that doesn't stop chomping caterpillars living on the leaves and guzzling grisly bits of dead insects.

8. **Jimson weed** poison attacks the nerves. Victims suffer thirst and their legs move oddly, and they see things that aren't there.

9. Thirty or forty **oleander** leaves contain enough poison to kill a horse – literally. It would have to be a hungry horse, since the lethal leaves taste vile.

10. **Henbane** poison causes madness, fits, shaking arms and legs, and violence. In the Middle Ages a bit of henbane gave German beer quite a kick.

AWFUL ANIMALS

For an animal, every day brings its dangers. Brainy bunny Mr Fluffy has agreed to give us a rabbit's eye view of life in the wild and even show us his photograph album. Uh-oh – the shrinking scientists don't look too happy…

BUT RABBITS DON'T HAVE PHOTO ALBUMS!

GRRR — THIS BOOK IS GETTING SILLY!

HE'S NOT AS BRAINY AS US!

OH, LIGHTEN UP, THIS BOOK IS SUPPOSED TO BE FUN!

THE HORRIBLE SCIENCE INTERVIEW

Mr Fluffy the rabbit tells it like it is...

Horrible Science: So what's it like being a rabbit?

Mr Fluffy: Well, like all animals, I just want a quiet life and all the carrots I can eat.

Me eating carrots.

Horrible Science: Er, I'm not sure about the carrots… Anything else on your mind?

Mr F: You bet — I don't want to get eaten. Trouble is, loads of creatures want to eat me.

Here's a photo of when I nearly got snatched by a hawk...

Here's when a snake sneaked up on me...

I must say it helps to have eyes on the side of my head so I can spot anything creeping up on me. And it's lucky I'm a fast runner!

And here's me being chased by a fox...

Horrible Science: Your life must be pretty dangerous...

Mr F: Well, it's even worse if I fall sick... And then there's the blood-sucking fleas that live in my fur.

Me feeling flea-bitten and unwell.

Horrible Science (scratching): Ouch — I've just met them! Is there anything more you'd like out of life, Mr Fluffy?

Mr F: I want to father lots of baby bunnies ... and I could do with a packet of flea powder.

And now for three facts that will turn you into an instant zoologist. Oops, silly me. I bet you don't know what I'm talking about…

A **zoologist** is an animal scientist.

A **habitat** is the place an animal lives in.

THREE ANIMAL-TASTIC FACTS
(don't turn the page until you've read them)…

1 Every animal in the world has a habitat, where it finds food and shelter.

2 Animals depend on one another…

Meat-eating animals depend on plant-eating animals for food. And plant-eaters depend on meat-eaters. By eating weak and sick rabbits, foxes ensure that only the strongest bunnies get

OOER!

FOX = MEAT-EATING ANIMAL.

to mate, and this breeds strong, healthy rabbit families. And by keeping rabbit numbers down, the foxes stop the rabbits from guzzling all the grass and starving to death.

3 Scientists describe the complicated feeding links between animals as a food web…

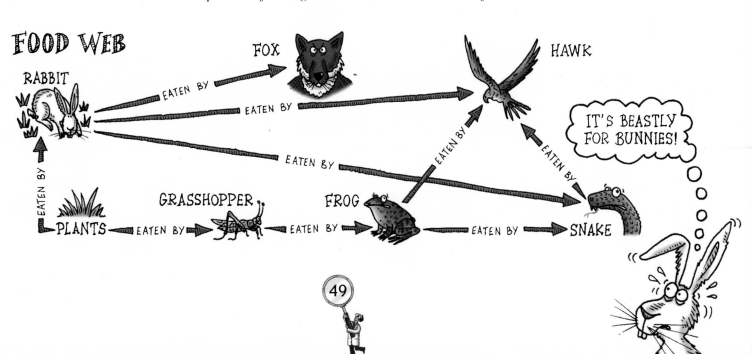

FOOD WEB

RABBIT — EATEN BY → FOX

EATEN BY → HAWK

IT'S BEASTLY FOR BUNNIES!

EATEN BY

PLANTS — EATEN BY → GRASSHOPPER — EATEN BY → FROG — EATEN BY → SNAKE

AWFUL ANIMAL LIFESTYLES

Every animal wants much the same things as Mr Fluffy – food, safety and babies. But, as you're about to find out, every animal has its own way of achieving these aims…

Finding food

Every type of food has its drawbacks…

• Plants are easy to catch (they don't run away) but you need to eat a lot of plants to feed yourself. The cellulose is tough to digest – and that means you need a big gut to eat it. Plant-eaters such as cows and rabbits have bacteria in their guts to rot plants so they can be digested. So does the hoatzin bird. This bulging Brazilian bird has a big stomach and spends its time eating and burping to get rid of the gas the bacteria make.

• Animal meat is easier to digest and contains more energy, but you've got to catch your dinner first.

And you need super-senses…

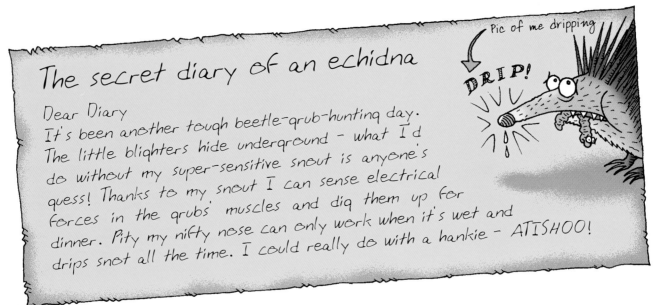

Staying safe

Animals try to hide or run from larger, fiercer animals, but if they get cornered, they have to fight back or act dead…

The secret diary of a hog-nosed snake

Dear Diary

Today was the worst day of my life! There I was happily snaking about when a dog leapt on me. Luckily I'm a good actor so I gave him my famous cobra impression – rearing up and hissing. It was quite a performance, if I say so myself, luvvies! But the miserable mutt just barked. Time for Plan B, I thought, and rolled over and played dead. Being dead is one of my best-ever acts, and to make it even more life-like, er, I mean death-like, I squirted smelly juice as if I was rotting. That foiled foolish Fido and he slunk off... I reckon it must be worth an Oscar!

Me doing my cobra thing →

Having kids

All animals have to produce young or their species will die out. There are two ways to make sure as many babies survive as possible…

• Produce lots of babies and *don't* look after them. The strongest will survive without help and having more babies means more should make it. Examples include fish and frogs.

• Produce a few babies but care for them. That way there's less chance of losing any. Examples include mammals such as humans.

But even animals that care for their young sometimes put them to work…

The secret diary of a naked mole rat

This is me (naked)

Dear Diary

I hate my life! I mean, it's bad enough looking like a sausage with fangs ... but Mum gives me and my brothers and sisters a hard time! She's always biting us and making us work all day in our tunnel garden. Today I had to chew roots and plant them in the soil so they will regrow. And afterwards I had to feed my baby brothers and sisters with my own poo. I wish I was a real rat – then I'd bite back!

Hopefully we humans aren't quite that bad, and in the next chapter we'll be taking a good hard look at ourselves – inside and out. (The size-sorting process lets us fit in next because we're bigger than most animals.) So read on, and see the stunning secrets of bloody body bits in glorious gory colour – if you're brave enough to keep your eyes open, that is!

PLIFFY SHPLIP LA ZIK SPLOG?*

VOMSAK

*ANYONE WANT A SICK BAG?

BLOODY BODY BITS

If your body was a machine, it would be the most marvellous machine in the universe. Just think – a machine fuelled on chewed-up plants and meat that also repairs itself! And that's just the start of the wonders of your brilliant bod, as you're about to discover…

We've sent the shrinking scientists on the world's most gruesome guided tour – *inside* the human body!

Horrible Holidays present...
THE BLOOD, BONES AND BODY BITS TOUR

Get shrunk down to the size of a pinhead and climb aboard our flying, swimming micro sub (sick bags provided)

- Gasp at the gurgling guts in disgusting digestive detail!
- Stop for lunch in the guts (if you're hungry, that is!)
- Take a breather in the windy lungs and heaving heart!
- Pause for thought in the brilliant brain!

YUCK! I wonder who would be money-grabbing enough to let people look around their *insides* for lots of cash?!

YOU'RE WELCOME TO GO TAKE A LOOK, BUT I WANT THE CASH UP FRONT!

The Blood, Bones and Body Bits Tour – Part 1

Welcome to the most gruesome guided body tour ever! Before we set off, please listen to the following safety announcement…

SAFETY ANNOUNCEMENT
The tour begins with a short stroll on this skin. Be careful not to skid in the bacteria. Just 1 square cm of skin can have 100,000 of the blighters!

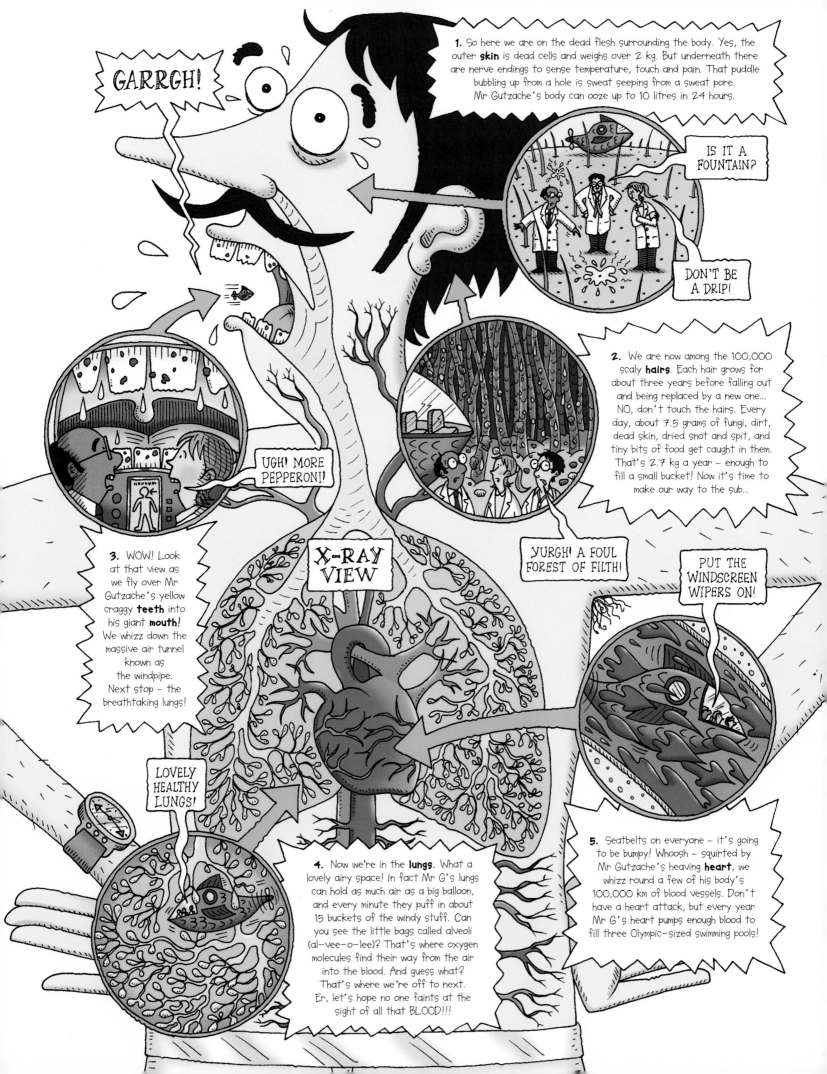

THE BLOOD, BONES AND BODY BITS TOUR – PART 2

Welcome to Part 2 of our gruesome guided tour of the body! On this part of the trip, we'll see what happens to Mr Gutzache's food. But we've picked his nose to start off with. Picked *what?* Oh yuck, you'd better wash your hands!

THE SUB'S SHAKING!

1. After returning to Mr Gutzache's **lungs** and getting puffed out up his windpipe, we find ourselves in the snotty cave behind his nose. The body sniffs food, and sensors in the roof of the cave trap airborne food molecules and send nerve signals to the brain... Much of what we call "taste" is actually the **smell** of food.

STOP SNIFFING, GUTZACHE!

2. Isn't the **tongue** a tasteful sight? Each bump samples food molecules dissolved in the watery spit. If the mouth is dry, food tastes like cardboard and cotton wool mixed together... Oops, watch out!

INCOMING ONION! DIVE! DIVE!

3. Here in Mr G's **bones** it looks peaceful compared with the boiling blood and thundering heart, but don't be bone-boozled. Inside the bones the marrow is busily churning out new blood cells. There are red blood cells that carry oxygen to the cells that make up the body, and white blood cells to fight bacteria. HELP – some white cells are after us!

4. Sorry about the bumpy ride! We're being churned in a bag of sick – the **stomach**. Let's use our sub's sensors to check out what Mr Gutzache's been eating...

OOER! THEY MUST THINK OUR SUB IS A MASSIVE MICROBE!

URGH! WELL THERE'S THE ONION FOR A START!

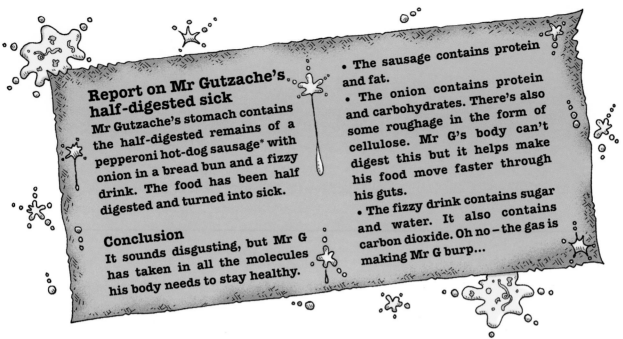

Report on Mr Gutzache's half-digested sick

Mr Gutzache's stomach contains the half-digested remains of a pepperoni hot-dog sausage* with onion in a bread bun and a fizzy drink. The food has been half digested and turned into sick.

Conclusion

It sounds disgusting, but Mr G has taken in all the molecules his body needs to stay healthy.

- The sausage contains protein and fat.
- The onion contains protein and carbohydrates. There's also some roughage in the form of cellulose. Mr G's body can't digest this but it helps make his food move faster through his guts.
- The fizzy drink contains sugar and water. It also contains carbon dioxide. Oh no – the gas is making Mr G burp...

*Goodness me – it's the hot dog from page 29!

HELP! The burps are stirring up a storm in the stomach. Er, I feel seasick – anyone got a sick bag? Oh dang! Tiny pits in the walls of the stomach are pumping out acid all around us to dissolve the food, and if we don't escape, we'll be dissolved too!

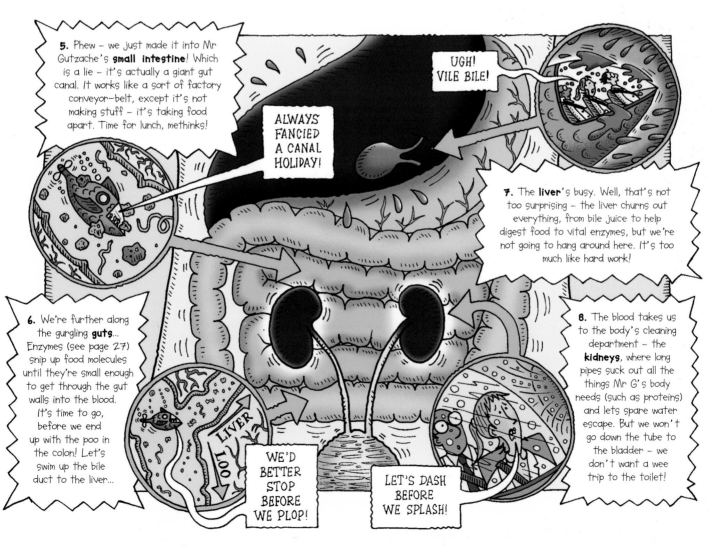

5. Phew – we just made it into Mr Gutzache's **small intestine**! Which is a lie – it's actually a giant gut canal. It works like a sort of factory conveyor-belt, except it's not making stuff – it's taking food apart. Time for lunch, methinks!

ALWAYS FANCIED A CANAL HOLIDAY!

UGH! VILE BILE!

7. The **liver**'s busy. Well, that's not too surprising – the liver churns out everything, from bile juice to help digest food to vital enzymes, but we're not going to hang around here. It's too much like hard work!

6. We're further along the gurgling **guts**... Enzymes (see page 27) snip up food molecules until they're small enough to get through the gut walls into the blood. It's time to go, before we end up with the poo in the colon! Let's swim up the bile duct to the liver...

8. The blood takes us to the body's cleaning department – the **kidneys**, where long pipes suck out all the things Mr G's body needs (such as proteins) and lets spare water escape. But we won't go down the tube to the bladder – we don't want a wee trip to the toilet!

WE'D BETTER STOP BEFORE WE PLOP!

LET'S DASH BEFORE WE SPLASH!

THE BLOOD, BONES AND BODY BITS TOUR – PART 3

We're in a nerve on our way to the thinking person's favourite body bit. That's right … Mr Gutzache's BRAIN. It's easy to get lost because nerves are like tangled telephone wires. Just like telephone wires, they carry messages - signals from the body to the brain and instructions from the brain to the muscles…

The massive muscles

OOPS, sorry folks – wrong turning! The muscle is a bundle of stringy thingies called fibres that shorten when the muscle pulls on a bone. By the way, no muscle can push – but at least they pull their weight, ha ha!

On our journey to Mr G's brain we'll take in the eyes and ears. This could be a real treat for our senses…

The awesome eyeballs

Welcome to the eyeball experience! This giant cave full of jelly-gloop has a lens at the front to focus light on to light-sensitive cells at the back. See what I mean?

The eerie ears

The ears are a complicated arrangement of eardrum, bones and sensitive hairs to pick up sound waves. These waves of movement in the air molecules trigger nerve signals to the brain. Er, keep your voices down or you'll deafen Mr Gutzache. Oh and don't get stuck in his gooey yellow earwax…

The brain, at last!

Here's the brain – what a mighty organ! It really does make you think, doesn't it?!

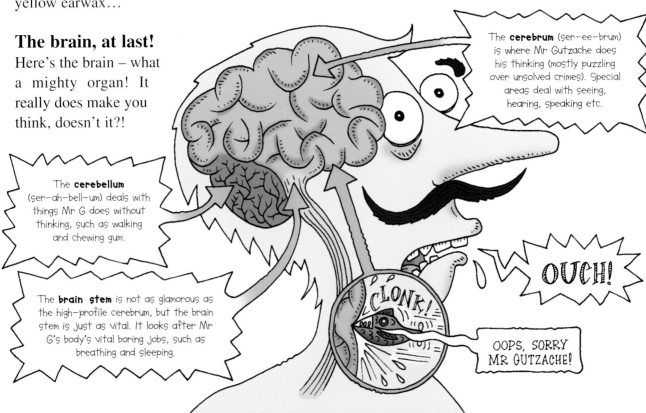

The **cerebrum** (ser-ee-brum) is where Mr Gutzache does his thinking (mostly puzzling over unsolved crimes). Special areas deal with seeing, hearing, speaking etc.

The **cerebellum** (ser-ah-bell-um) deals with things Mr G does without thinking, such as walking and chewing gum.

The **brain stem** is not as glamorous as the high-profile cerebrum, but the brain stem is just as vital. It looks after Mr G's body's vital boring jobs, such as breathing and sleeping.

OUCH!

OOPS, SORRY MR GUTZACHE!

Sadly that's the end of our tour, so let's head back in the blood to the lungs, and then get coughed out of the body...

YIPPEE!

COUGH!

CONGRATULATIONS! You've survived the Blood, Bones and Body Bits Tour and you saw some breathtaking body bits...

Bet you never knew!
Brainy scientists can't get their brains around how the brain thinks and remembers. But they reckon thoughts might be made by brain cells firing signals to each other in the cerebrum. Since there are 60,000 brain cells in a squitty 1-mm-square bit of brain and TEN BILLION possible connections, it's easy to get lost in thought...

But you never saw what the body is *really* made of. Your body has more than 50 million million (50,000,000,000,000) cells. And you can't live without them. Let's take a really close peek at a cell. Aha – there's one!

Another embarrassing apology from the author...
Sorry, readers! There's another fault with the size-sorting process, and we're looking at something smaller instead of finding out about bigger things! Normal service will be resumed in the next chapter – I promise!

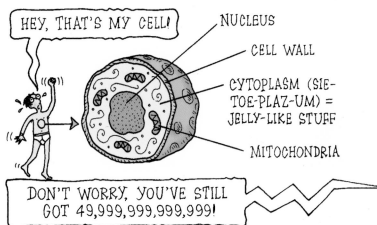

HEY, THAT'S MY CELL!

NUCLEUS

CELL WALL

CYTOPLASM (SIE-TOE-PLAZ-UM) = JELLY-LIKE STUFF

MITOCHONDRIA

DON'T WORRY, YOU'VE STILL GOT 49,999,999,999,999!

Notice anything? Yes, it's not too different to the protist on page 29 – and it's just as complicated. Inside the cell, there are over 20,000 different types of protein molecules – perhaps 100 million in all. I wonder who counted them?

But the most amazing secret lies inside the nucleus. That amazing DNA stuff is actually a secret code – but what does it mean? Turn the page and let's get code-cracking...

DNA CODE BOOK

PULL!

SWEAT! WHEEZE! GASP!

THE TOP-SECRET DNA CODE BOOK

Remember DNA? It's actually short for deoxyribonucleic acid (that's dee-ox-ee-rye-bow-new-klay-ick acid). DNA turned up in microbes and plants – and guess what? Animals and humans have it too. Anyway, this marvellous molecule is the body's secret code for making a brand-new body, and the shrinking scientists are on a mission to find out how it works…

The shrinking scientists in … clever code-crackers

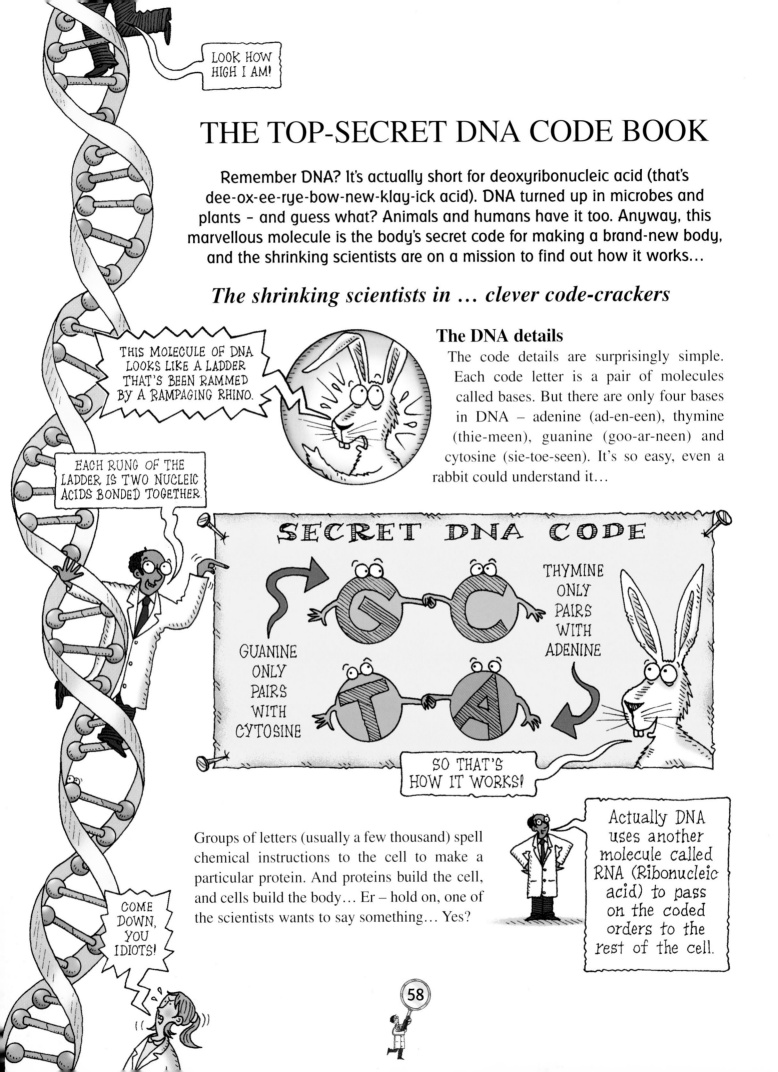

The DNA details

The code details are surprisingly simple. Each code letter is a pair of molecules called bases. But there are only four bases in DNA – adenine (ad-en-een), thymine (thie-meen), guanine (goo-ar-neen) and cytosine (sie-toe-seen). It's so easy, even a rabbit could understand it…

SECRET DNA CODE

GUANINE ONLY PAIRS WITH CYTOSINE

THYMINE ONLY PAIRS WITH ADENINE

Groups of letters (usually a few thousand) spell chemical instructions to the cell to make a particular protein. And proteins build the cell, and cells build the body… Er – hold on, one of the scientists wants to say something… Yes?

Speech bubbles:
- LOOK HOW HIGH I AM!
- THIS MOLECULE OF DNA LOOKS LIKE A LADDER THAT'S BEEN RAMMED BY A RAMPAGING RHINO.
- EACH RUNG OF THE LADDER IS TWO NUCLEIC ACIDS BONDED TOGETHER.
- SO THAT'S HOW IT WORKS!
- COME DOWN, YOU IDIOTS!
- Actually DNA uses another molecule called RNA (Ribonucleic acid) to pass on the coded orders to the rest of the cell.

Isn't that typical? The more science you learn, the more complicated it gets! Mind you, even the basic info about DNA is complicated enough…

SIX CRUCIAL COMPLICATED DNA FACTS

1 Most of your cells contain 1.8 metres of DNA. You've got enough DNA in you to stretch to the Sun and back over 400 times!

2 Each DNA molecule is folded into a blobby thingie called a chromosome (crow-ma-soam) containing 3.2 billion letters of code – enough to fill 5,000 books like this one (but I bet they're a lot more boring!)

3 Your DNA letters can be combined in 10 multiplied by 3.5 billion ways. So your chance of meeting a stranger with the same code as you is about 1 in 10 (add another 3.5 billion zeros here). And that's not too likely.

4 The codes that make a body feature such as your eye colour are called genes. (And you don't have to be a gene genius to realize they're nothing to do with blue denim.)

5 You'll be delighted to read that about 30,000 genes are needed to make that wonderfully complicated creature called YOU, and less happy to find out that this puts you on a level with grass.

6 And what's more, scientists aren't too sure what 97% of your DNA code actually does. In the past it's been called "junk DNA"…

Ever wondered where your DNA (including the junk) comes from? Well, here's a handy diagram to figure it out…

Ah yes, I was planning to tell you in the next chapter. The trouble is, you might get killed reading it… Rumour has it that hungry dinosaurs are roaming the next two pages, and they're none too friendly!

DREADFUL DINOSAURS
(AND OTHER FEARSOME FOSSILS)

Is BIG really better? Not necessarily! The dinosaurs were bigger than you, but I bet they weren't as brainy. In fact, I'm certain those dim-witted dinosaurs couldn't even read this book! Anyway, this chapter's all about dinosaurs and other prowlers from the past, so let's visit our very own Horrible Science dinosaur park…

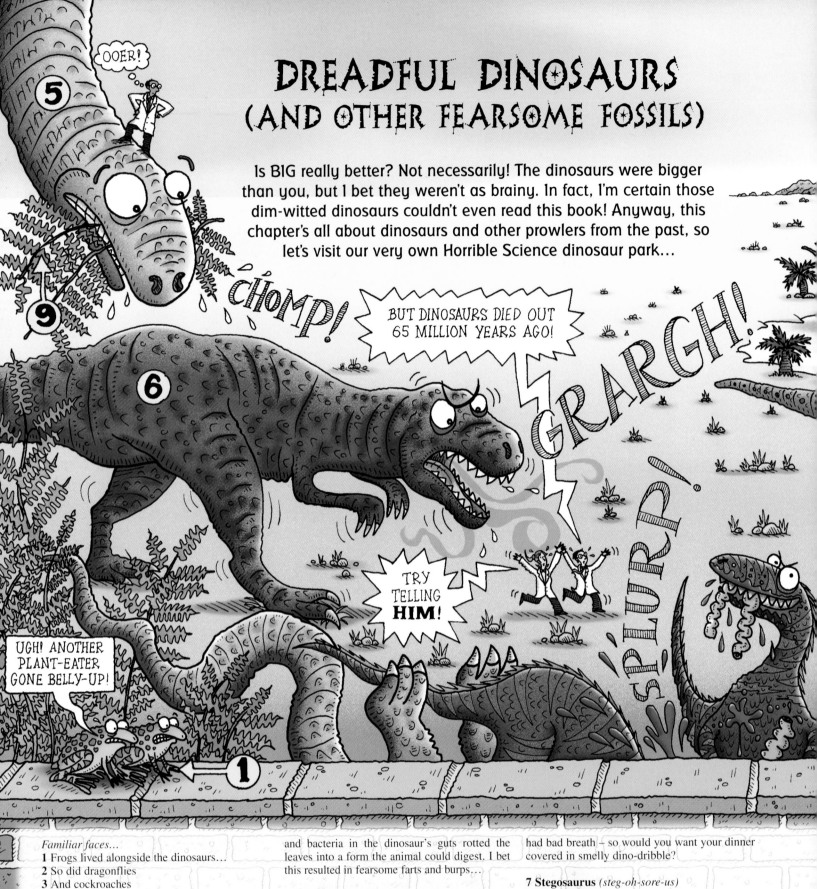

OOER!

CHOMP!

BUT DINOSAURS DIED OUT 65 MILLION YEARS AGO!

GRARGH!

SPLURP!

TRY TELLING **HIM!**

UGH! ANOTHER PLANT-EATER GONE BELLY-UP!

Familiar faces…
1 Frogs lived alongside the dinosaurs…
2 So did dragonflies
3 And cockroaches
4 And even crocodiles!

5 Barosaurus *(ba-row-sore-us)*
SIZE: 27 metres long (9-metre neck and 13-metre tail). Dinosaurs with long necks and tails were called **Sauropods** *(sore-ro-pods)*.
LIVED: 150 million years ago
ATE: Leaves from trees (it could rear up on its hind legs)
Dreadful dino-detail: Barosaurus guzzled leaves,

and bacteria in the dinosaur's guts rotted the leaves into a form the animal could digest. I bet this resulted in fearsome farts and burps…

6 Tyrannosaurus rex *(tie-ran-oh-sore-us rex)*
SIZE: 12 metres long from its giant jaws to its terrible tail
LIVED: 65 million years ago
ATE: Plant-eating dinosaurs and any dead dinosaurs it could sink its gigantic jaws into
Dreadful dino-detail: T. rex fed by tearing chunks of flesh from the bones and twisting its nasty neck. Each bite of meat could feed a human family for a month. But the dreadful dinosaur

had bad breath – so would you want your dinner covered in smelly dino-dribble?

7 Stegosaurus *(steg-oh-sore-us)*
SIZE: 9 metres long
LIVED: 150 million years ago
ATE: Juicy plants
Dreadful dino-detail: A knot of nerves in its back probably controlled the creature's legs. This knot was bigger than the stupid Stegosaurus's nut-sized brain.

8 There were lots of volcanoes towards the end of the dinosaur era.

9 Grass hadn't appeared at the time of the dinosaurs. But there were plenty of ferns…
10 And pine-trees to munch.
11 Flowering plants appeared about 100 million years ago. In the 1930s scientist Ernest Baldwin suggested that plants containing laxative oils became rarer, and plant-eating dinosaurs died of constipation. Meat-eaters then starved. So I guess the dinosaurs were "dung to death".

In the age of dinosaurs, mammals probably spent their lives hiding from the dreaded dinosaurs. Can you spot the **Megazostrodon** (*meg-ah-zoh-strow-don*) in this picture?

12 Deinonychus (*die-noh-nye-cuss*)
SIZE: 3 metres long
LIVED: 113 million years ago
ATE: Hunted in packs and leapt on large plant-eating dinosaurs
Dreadful dino-details: Dreadful Deino probably had feathers. It ripped open its victim's belly with its terrible toe claws and gobbled the grisly guts. Anyone for *sausages*?

13 Pachycephalosaurus (*pak-ee-sef-a-low-sore-us*)
SIZE: 5 metres long (guess based on size of skull)
LIVED: 68–65 million years ago
ATE: Plants and possibly dead animals

Dreadful dino-detail: The name means "thick-headed lizard" and the males probably had butting contests over females at breeding time. You'd have to be a real thickhead to try this…

14 Triceratops (*Try-ser-ah-tops*)
SIZE: 9 metres long
LIVED: 67–65 million years ago
ATE: Tough plants
Dreadful dino-detail: Triceratops teeth have turned up in rock dating from *after* the dinosaurs died out. The teeth may have been washed out from older rocks. Either that or someone stole a Triceratops' false teeth…

DEADLY DINOSAUR SCIENCE

Dinosaurs are great, but the scary science behind them is even more stunning. To find out more we'll be taking a close look at a hungry T. rex, but first let's meet the brainy boffin who invented the name "dinosaur"…

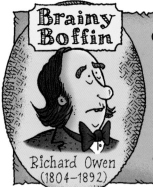

Brainy Boffin

Richard Owen (1804–1892)

Owen was the scientist who gave dinosaurs the WRONG name. You see the word dinosaur means "terrible lizard" in Greek, but as you're about to find out, dinosaurs weren't lizards at all! As a young man, revolting Richard trained as a doctor, and his favourite hobby was cutting up dead bodies. One day he dropped a man's head in the street and it rolled into a cottage… And I guess that's when he decided it was safer to study ancient animal bones.

And talking about ancient bones, have you ever wondered what a fossil is? No, it's not another word for your prehistoric science teacher!

A **fossil** is the trace of a living thing preserved in rock. When dinosaur bones rotted, minerals could fill the space left and form fossil bones.

FASCINATING FOSSIL FACTS

1 To make a fossil, the bones have to be covered by mud or sand before they rot away. This is quite rare – in fact 99.9 per cent of dinosaurs never became fossils.

2 Scientists have found the fossils of about 250,000 different plant and animal species. It sounds DEAD impressive until you learn that more than FIVE BILLION species might have lived on Earth during its history. And they've nearly all disappeared without a trace!

3 Mind you, it's amazing what can turn up as a fossil. Scientists have found dinosaur eggs, dinosaur poo and even fossil sick from an ancient sea reptile… I guess it must have been a sicky-saur.

Welcome back to the Horrible Science lab where we're just giving T. rex a rex-ray, sorry X-ray…
Er, is this really such a good idea?

EVOLUTION MADE EASY

Evolution explains how animals change over thousands of years. In a moment we'll meet the brilliant barnacle-brained boffin who figured it out – but first let's answer that question from page 59 about where DNA comes from. We've invited the readers back to see their family tree…

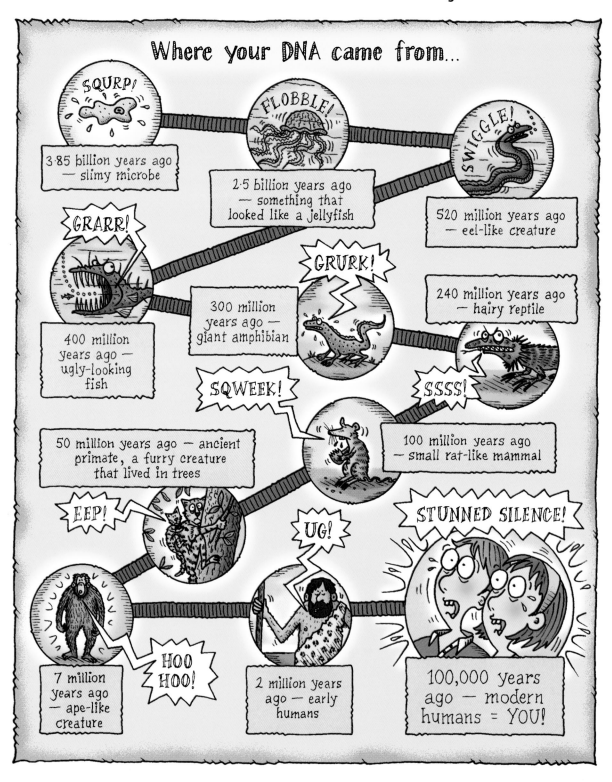

And here are the full fascinating facts about how all these creatures evolved over time and ended up as you…

STUNNING SCIENCE FACT FILE
Name: How evolution works

THE BASIC FACTS:
• DNA is the code that controls your appearance, remember?
• Every animal has its own special DNA code that it passes on to its young.
• The DNA code can be changed (for example, by damaging chemicals). Other changes can happen when DNA is copied to make new cells. Changes to the code make the offspring look different.
• For an animal, every day is a battle to find food and avoid being food for another beast (if you don't believe me, just ask Mr Fluffy).

Animals with features that help them in this battle are more likely to breed and pass on these advantages in their DNA.
• Over time all the animals in this species will take on the new, improved features…

NEW IMPROVED EARS!

THE STUNNING DETAILS:
If just one of all these creatures that were your ancestors hadn't mated and passed on their DNA, you wouldn't exist today!

Hmm, what an incredible idea! I bet you'd just love to meet the gobsmacking genius who thought it up?

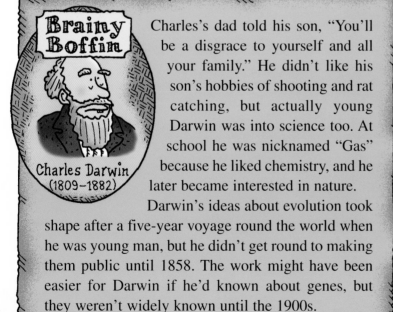

Brainy Boffin

Charles Darwin
(1809–1882)

Charles's dad told his son, "You'll be a disgrace to yourself and all your family." He didn't like his son's hobbies of shooting and rat catching, but actually young Darwin was into science too. At school he was nicknamed "Gas" because he liked chemistry, and he later became interested in nature.

Darwin's ideas about evolution took shape after a five-year voyage round the world when he was young man, but he didn't get round to making them public until 1858. The work might have been easier for Darwin if he'd known about genes, but they weren't widely known until the 1900s.

Bet you never knew!
Before he wrote his book on evolution, Darwin spent eight years studying barnacles. He kept 10,000 of them in tanks at home. I bet they got in the way and his wife told him off…

YOU'RE SO "SHELLFISH"!

CLONK!

D-DAY FOR DINOSAURS

And now for a scary thought – 99.9 per cent of all species that have ever lived are extinct. Sixty-five million years ago it was the dinosaurs' turn. Scientists think Earth was whacked by an asteroid several kilometres across.

An **asteroid** is a lump of space rock.

Extinction is when a species dies out.

Here's how a young Pachycephalosaurus might have remembered the disaster…

65 million years ago
North America

Dear Mum,
What a terrible day! Me and the herd were snacking on leaves when a light streaked across the sky brighter than the Sun! One second later came a super-hot blast with the force of seven billion atom bombs. The sky went black and we were blown off our feet. No sooner had I got up when I was bashed on the head by burning rocks. (It sometimes helps to be a thickhead!) Anyway that was this morning but even now it's raining acid, and umbrellas haven't been invented yet! The weather forecast is cold for 10,000 years, plus lots of choking gas from volcanoes... Now call me a thick-headed lizard, but something tells me things aren't looking too good for us dinosaurs. Is this the end of the AAAARRRRRGGGGGH!

But the end of the dinosaurs was good news for US. During the dinosaur era, your ancestors were tiny, timid and terrified. But after the asteroid did for the dinosaurs the mammals bounded from their burrows and took over the Earth. Yes, if it wasn't for that lovely lump of space rock, you could be 5 cm long and live in a hole…

After the dinosaurs a whole host of curious creatures evolved…
1 Titanis was a 2.5-metre-tall fearsome flesh-eating bird that lived 50 million years ago. It sounds nasty enough to give your pet cat personality problems and I bet it had a fowl temper.
2 Imagine a guinea pig as big as a bad-tempered buffalo. In 2002 scientists found the remains of this creature in Venezuela. Just be grateful you don't have to clean out its cage.
3 The Paraceratherium (pa-ra-sura-thee-ree-um) of 35 million years ago was a sort of rhino the size of a house. At 8 metres tall it was the largest land mammal ever.
4 The Australian Procoptodon (pro-kop-toe-don) was a 3-metre-tall killer kangeroo. If it caught you on the hop, you'd be in for the high jump.

OOER – WHO RUFFLED HIS FEATHERS?

And so, at last…
Seven million years ago, in Africa, an obscure primate stood on its hind legs and learnt how to make tools. Slowly it became smarter. It evolved into new species that learnt how to use fire and wear clothes and grow crops and read Horrible Science books and become YOU. Now that's what I call progress! In 2001 African scientist Ahounta Djimdoumalbaye discovered the remains of a prehistoric primate that lived in east Africa over six million years ago. This could have been one of our most ancient ancestors.

So here we are. Right now we rule the planet and we're doing OK. But could we go the way of the dinosaurs? In the long term things don't look too good…

Bet you never knew!
Extinctions can happen in splurges. The disaster that did in the dinosaurs also bumped off 70 per cent of all species on Earth. And 250 million years ago something very nasty killed off 96 per cent of species! Some scientists think giant volcanoes burped vast amounts of gases like carbon dioxide, changing the chemistry of the seas and killing off its creatures. Gasp!

Mind you, our pretty blue planet has lots of hidden horrible habits. We've sent a shrinking scientist into the boiling heart of the next chapter to find out more…

TURN BACK BEFORE IT'S TOO LATE!

THE EXPLOSIVE EARTH

Our journey from the terribly tiny to the horribly huge has at last landed us on something breathtakingly big – Planet Earth. In this down-to-earth chapter we'll explore bits of Earth you never knew existed and drop Mr Fluffy down a very deep hole. But first, let's grab a giant knife and fork and cut a huge slice out of the planet...

Planet Earth at a glance...

Diameter (distance across):
From east to west – 12,713 km
From north to south – 12,756 km

Circumference (distance around):
From north to south pole and back again the other side –
40,000 km
Round the middle (this line is called the equator) –
40,075 km

HOLD IT STEADY, WILL YOU?

Weight: a bit less than **6,000 million million million tonnes**
Age: **4.6 billion years**

YIKES! A 12,000-KM CLIFF!

Crust, 7-70 km of rock

Mantle, 2,900-km-thick hot rocks and churning currents make the plates move. (And hot means HOT – it's 1,500-3,000°C.)

Outer core, 2,200 km of melted iron and nickel – at 4,500°C it's even hotter. Churning currents of hot metal make the Earth's magnetic force.

Inner core, 2,500 km across and 5,500°C. Scientists think it's solid iron and nickel, but no one's totally sure because no one's been down to check it out ... any volunteers?

STUNNING!

YEAH – IT'S GROUND-BREAKING!

The Earth's surface is made up of plates – a bit like a broken egg-shell. There are eight big plates and about ten smaller ones. They're up to 40 km deep.

Bet you never knew!

When plates push together they make mountains, and that's handy because without mountains we'd be 9 km under the sea and fish would rule the world.

Thermosphere – above mesosphere – ultraviolet rays from Sun whack into air molecules and heat them to 1,500° C!

Mesosphere – 50-80 km. Brr – it's -90°C!

Stratosphere – 12-50 km – starts with tops of highest clouds. You could get there in 20 minutes in a lift (if there was one) but you'd best wrap up warm – it's -57°C!

Troposphere – 12 km thick. Weather happens here. Air molecules trap heat and keep us warm and cosy.

70 per cent of the Earth is covered by sea. Deepest ocean = Pacific. On average, it's 4.19 km deep.

The atmosphere (air around the Earth) is 500 km thick. Shrink the Earth to the size of a beach ball and the atmosphere would be no thicker than three layers of paint.

Terrible thought experiment – rabbit in a hole

ERK!

NORTH POLE – I MEAN HOLE

Imagine digging a hole right through the Earth. Now take Mr Fluffy up in a plane and drop him down the hole. (Please don't phone in to complain – it's only a thought experiment!)

People on the other side of the Earth would see the rabbit whizz out of the hole and fly almost as high as the original drop.

Then he'd fall back down the hole and fly out of the hole on your side of the Earth, almost as high as he started.

EEK, I'M FALLING...

...NOW I'M RISING!

After whizzing backwards and forwards a few times Mr Fluffy ends up floating at the centre of the Earth.

Most of Earth's gravity comes from the mantle and core, and that means at the centre of the Earth the force of gravity is equal in all directions. That means Mr Fluffy is weightless. Er – we'd better rescue him before he gets cooked!

CORE! IT'S HOT DOWN HERE!

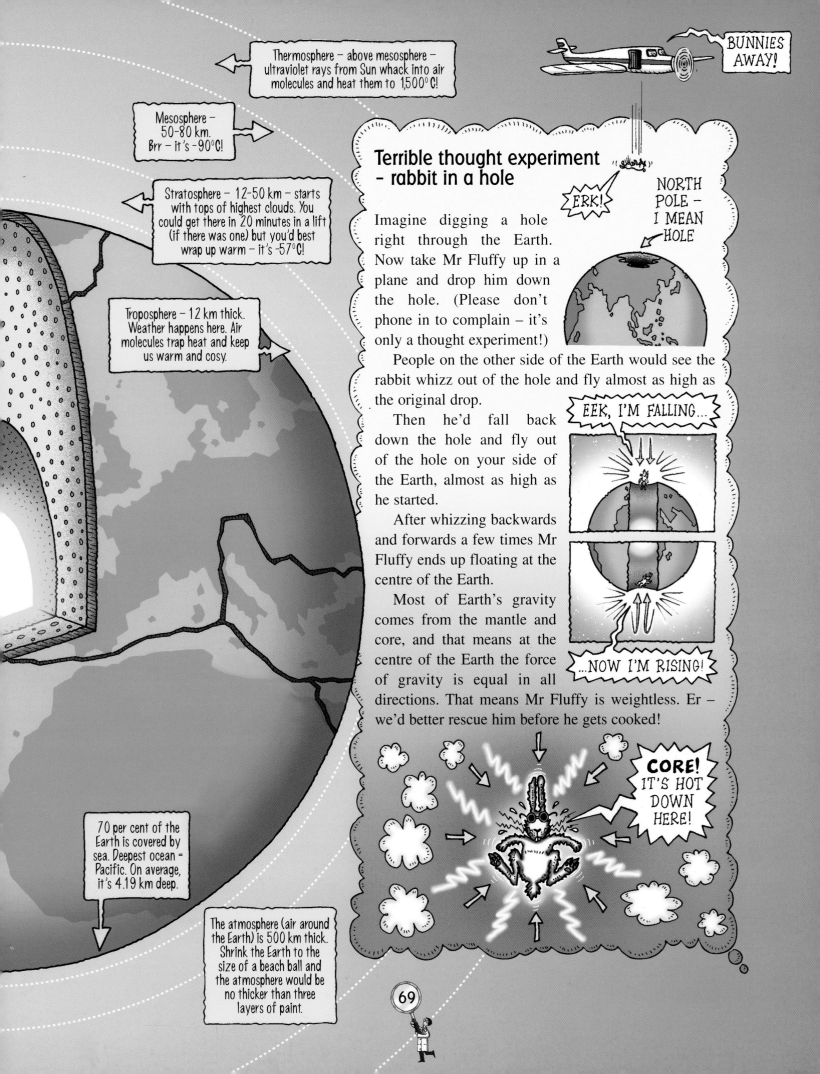

DESPERATE MEASURES FOR EARTH

It's easy to check out the size and shape of Earth in this book, but hundreds of years ago scientists weren't sure of these facts. And the suffering scientists made extreme efforts to find out…

In 1735 Pierre Bouguer, Charles de la Condamine and a group of French scientists travelled to Ecuador, South America, to work out the exact shape of the Earth. Calculations of Earth's gravity suggested that the planet was bulging around the middle and a little flatter at the poles – but were they right? Well, here's what Bouguer's diary might have looked like…

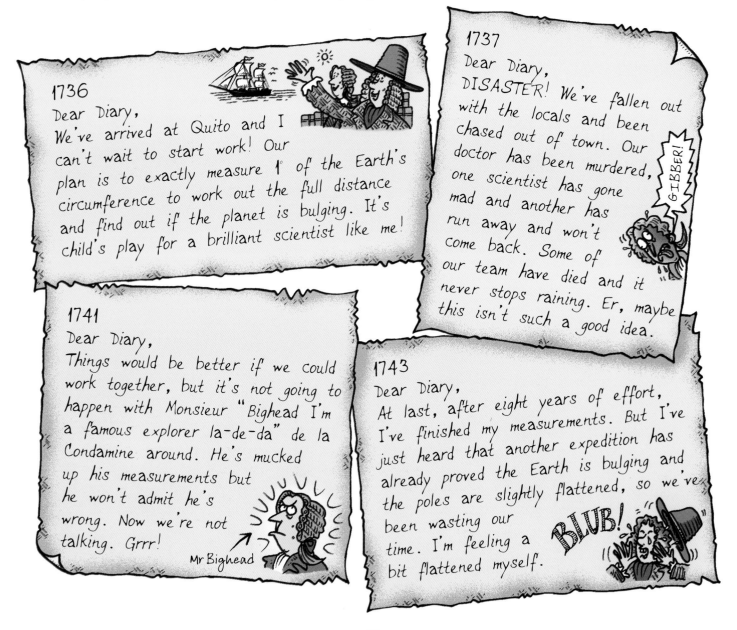

1736
Dear Diary,
We've arrived at Quito and I can't wait to start work! Our plan is to exactly measure 1° of the Earth's circumference to work out the full distance and find out if the planet is bulging. It's child's play for a brilliant scientist like me!

1737
Dear Diary,
DISASTER! We've fallen out with the locals and been chased out of town. Our doctor has been murdered, one scientist has gone mad and another has run away and won't come back. Some of our team have died and it never stops raining. Er, maybe this isn't such a good idea.

GIBBER!

1741
Dear Diary,
Things would be better if we could work together, but it's not going to happen with Monsieur "Bighead I'm a famous explorer la-de-da" de la Condamine around. He's mucked up his measurements but he won't admit he's wrong. Now we're not talking. Grrr!

Mr Bighead

1743
Dear Diary,
At last, after eight years of effort, I've finished my measurements. But I've just heard that another expedition has already proved the Earth is bulging and the poles are slightly flattened, so we've been wasting our time. I'm feeling a bit flattened myself.

BLUB!

You'll be happy to hear that Bouguer and Condamine got back to France safely. Even if they did travel on separate ships.

A WEIGHTY question…

Meanwhile scientists were busily trying to weigh the Earth. In 1774 astronomer Nevil Maskelyne (1732–1811) spent a summer in a tent beside a Scottish mountain getting rained on and bitten by bugs.

Here's what he was trying to do…

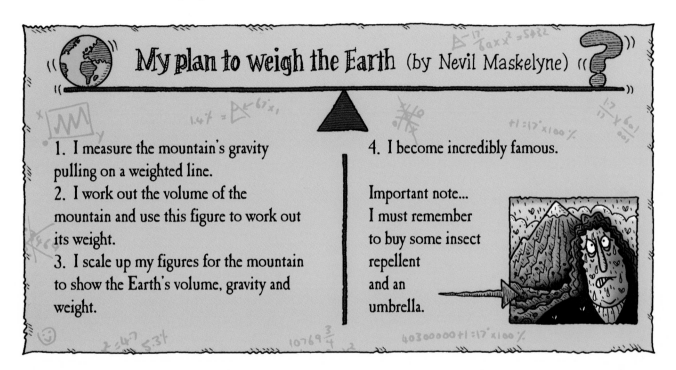

My plan to weigh the Earth (by Nevil Maskelyne)

1. I measure the mountain's gravity pulling on a weighted line.
2. I work out the volume of the mountain and use this figure to work out its weight.
3. I scale up my figures for the mountain to show the Earth's volume, gravity and weight.

4. I become incredibly famous.

Important note…
I must remember to buy some insect repellent and an umbrella.

Ultimately, all the suffering and maths was in vain. A fire burnt down Maskelyne's tent (and a local fiddler's violin). And the final figure for the Earth's weight was WRONG by 1.6 thousand million million million tonnes.

Brainy Boffin

Henry Cavendish. (1731–1810)

In 1794 this very strange scientist weighed the Earth more accurately without leaving home. Crazy Cavendish only ever talked about science, and if he didn't want to talk, he'd squeak and run away. Anyway, he measured the gravitational pull between large and small metal balls, and used the figures for their weight and gravity to show the Earth weighed 6,000 million million million tonnes.

Cavendish was almost spot on. We asked him how he felt…

SQUEAK!

WELCOME TO THE GREATEST QUIZ SHOW ON EARTH!

HOW WELL DO YOU KNOW YOUR PLANET?

IF YOU GET THE QUESTIONS RIGHT YOU'LL WIN YOUR VERY OWN READY-TO-MOVE-INTO PLANET!!!!

GASP!

HA HA! ONLY KIDDING!

GROAN!

Round 1

Where in the world?
Where on Earth are the following features?
1 The world's highest waterfall.
2 The world's longest mountain range.
3 The oldest rock on Earth.

c) The North Pole
b) Denmark Strait
a) Himalayas
g) Africa
d) Australia
e) Hawaii
f) The middle of the Atlantic, Indian and Pacific Oceans

Round 1 answers:
1 b) Yes, the highest waterfall! The Denmark Strait is a 3.5-km-high waterfall, where cold water from the northern seas sinks (it's heavier than the warmer water it meets). The cold water forms an underwater river 1,000 km long.

2 f) The massive mountain range stretches the length of these oceans and marks the site where plates are pushing apart and hot rock is flooding to the surface. Hawaii, the Azores and the Canary Islands are extra-high mountains.

3 d) Yes, it's 4.3 billion years old. By the way, if you're wondering how scientists know how old a rock is without asking it, it's all to do with the amount of uranium it has (remember that element from page 17?). Uranium atoms fall to bits and turn to lead at a regular rate – which scientists measure. S'easy when you know how!

Wet, wet, wet ... and more wet!

Here's what happens to a molecule of water on planet Earth. But which of these facts are true and which are too wet for words? **TRUE or FALSE?**

1 A drop of water can spend 1,000 years underground.

2 If all the rain in all the clouds got caught in buckets, you'd fill enough buckets to stretch to planet Neptune ... and back.

3 It's true that water wears away rock – it's called erosion. But is it also true that the Niagara Falls are eroding the rocks they rush over by 1.1 metres every thousand years?

4 A fresh supply of water falls from space every day.

WATER TURNS INTO VAPOUR

WATER VIEW!

RAIN CLOUD

PLANTS LOSE WATER INTO AIR

WATER FLOWS OUT TO SEA

WATER SOAKS INTO GROUND

SPECIAL BONUS QUESTION

The bottom of the oceans are covered in SMILE. But if you got stuck in it you wouldn't be smiling! Simply rearrange the letters to get the right answer and win **TWO POINTS!**

Round 2 answers:

1 TRUE If water falls on plants it escapes from their leaves in a few hours. If it trickles down into rock it's stuck for even longer than the average science lesson.

2 TRUE Every day enough rain falls to give everyone on Earth 900 baths. I reckon that many baths would drive you *clean* out of your mind.

3 FALSE It's 1.1 km. And that means in 25,000 years' time the Falls won't exist because they'll have eaten their entire river.

4 FALSE All the water on Earth has been swilling around for 3.8 billion years – scientists think it arrived as ice in comets. Chances are the water you drank today had been dribbled by dinosaurs, farted in by frogs and peed in by hundreds of people.

Bonus question answer:

SLIME. The ocean ooze is made up of dead plants and animals, and builds up 6 metres every million years.

So here's what your score means...

0–3 *You're not from round here are you? You're not the purple slobwobbler by any chance?*

4–7 *You've got a good DOWN-TO-EARTH knowledge of your planet.*

8+ *ON TOP OF THE WORLD!!! Congratulations – you'll make a really stunning Earth scientist!*

HOW TO DESTROY THE EARTH
(THE MOVIE)

How do you feel? Safe, sleepy, snug? Don't get too comfy! If these disaster movies are to be believed, at any moment the Earth could blow up or turn into a giant snowball. And I bet that could spoil your whole day…

Hmm – I wonder what the shrinking scientists make of this danger?

So are we in for the BIG CHILL-OUT? We asked the shrinking scientists…

Some scientists think the Earth actually was a giant snowball 750 million years ago. They say heavy rains washed most of the carbon dioxide from the air, and without the gas to trap heat Earth cooled and froze.

At present the climate seems to be warming, but there could be an ice age in the next few thousand years…

And I'm ready for it!

STUNNING SCIENCE FACT FILE
Name: Ice ages

THE BASIC FACTS:
• In an ice age, ice up to 3 km thick covers the northern parts of the Earth. It's not as bad as 750 million years ago, but you wouldn't want to leave home without your woollies.
• Ice ages could be triggered by slight changes in the Earth's orbit (path) round the Sun, cutting down the amount of sunlight we get. Ice builds up and reflects away more sunlight – cooling the planet.

THE STUNNING DETAILS:
If the world's ice melted tomorrow, the sea level would rise to the height of a 20-storey skyscraper. Fancy a swim?

WOOLLY MAMMOTH FROM LAST ICE AGE

MAMMOTH WOOLLY FOR NEXT ICE AGE

TWO REASONS WHY YOU MUSTN'T PANIC WHEN IT SNOWS
1 Our Earth is a giant heater. Deep underground, zappy radioactive uranium gives off heat that gets trapped deep below the surface. And there's lots of heat left from when the Earth was formed and giant rocks were crushed together.
2 Remember how Melanie set off a volcano to save the world? Well, scientists who believe Earth was a giant snowball believe volcanoes blasted out enough carbon dioxide to trap heat and warm the planet again.

SWEAT!

Well, that's nice! So we're not going to get blown up or deep-frozen anytime soon? No – the real risk is being clunked on the bonce by a space rock, as happened to the died-out dinosaurs. And there's loads of space rocks whizzing about in the next chapter, together with putrid planets and the sizzling Sun. Ready to blast off?

IT'S A BIG PLACE...

THE SCARY SOLAR SYSTEM

ASTRONOMICALLY BIG!

Once again, this chapter's bigger than the last, and this time we're off to see the Solar System – that's the Sun and its surrounding planets. And we'll be seeking out a stack of stunning Solar System secrets to make you starry-eyed…

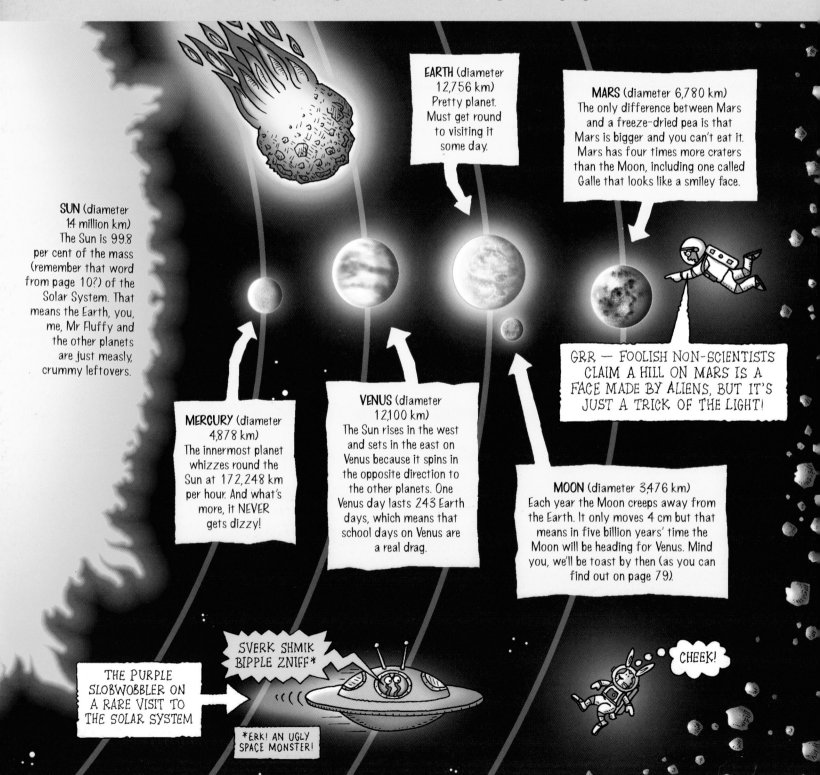

EARTH (diameter 12,756 km) Pretty planet. Must get round to visiting it some day.

MARS (diameter 6,780 km) The only difference between Mars and a freeze-dried pea is that Mars is bigger and you can't eat it. Mars has four times more craters than the Moon, including one called Galle that looks like a smiley face.

SUN (diameter 14 million km) The Sun is 99.8 per cent of the mass (remember that word from page 10?) of the Solar System. That means the Earth, you, me, Mr Fluffy and the other planets are just measly, crummy leftovers.

GRR — FOOLISH NON-SCIENTISTS CLAIM A HILL ON MARS IS A FACE MADE BY ALIENS, BUT IT'S JUST A TRICK OF THE LIGHT!

MERCURY (diameter 4,878 km) The innermost planet whizzes round the Sun at 172,248 km per hour. And what's more, it NEVER gets dizzy!

VENUS (diameter 12,100 km) The Sun rises in the west and sets in the east on Venus because it spins in the opposite direction to the other planets. One Venus day lasts 243 Earth days, which means that school days on Venus are a real drag.

MOON (diameter 3,476 km) Each year the Moon creeps away from the Earth. It only moves 4 cm but that means in five billion years' time the Moon will be heading for Venus. Mind you, we'll be toast by then (as you can find out on page 79).

SVERK SHMIK BIPPLE ZNIFF*

THE PURPLE SLOBWOBBLER ON A RARE VISIT TO THE SOLAR SYSTEM

*ERK! AN UGLY SPACE MONSTER!

CHEEK!

HERE COMES THE SUN!

We've decided to build a nice new Sun (complete with Solar System) to show you how the old one formed. It's a stunningly expensive project so any donations are welcome. Here's the plan…

OUR PLAN TO BUILD A NEW SUN AND SOLAR SYSTEM

SWIRL! + K-BOOM!

PULL PULL PULL PULL PULL PULL PULL PULL

SHINE!

1 We collect a huge swirling cloud of gas and dust 24 billion km across and blast it with a giant exploding star.

2 Tiny specks of dust start to clump together…

3 Within 200 million years, gravity will crunch the matter in the centre of the swirl tightly together to make the Sun, and we'll even have a lovely little family of planets going round it!

Likely cost: £10,000,456,234,998.22p

A quick note from the author…
1. This is how the Solar System formed 4.6 billion years ago.
2. On second thoughts, building a new Sun isn't such a HOT idea. It's a bit too expensive and you wouldn't want to wait 200 million years to finish this book, would you?

WHAT WAS WRONG WITH THE OLD ONE, ANYWAY?!

So instead we've sent the shrinking scientists and Mr Fluffy to take a close-up X-ray photo of the Sun.

HORRIBLE HEALTH WARNING!

You're far too sensible to look directly at the Sun or try to photograph it — aren't you? Just checking!

A quick science note…
The Sun's a huge, horribly hot ball of hydrogen and helium plasma (remember that word from page 20?) Now read on…

The shrinking scientists in … no fun in the Sun

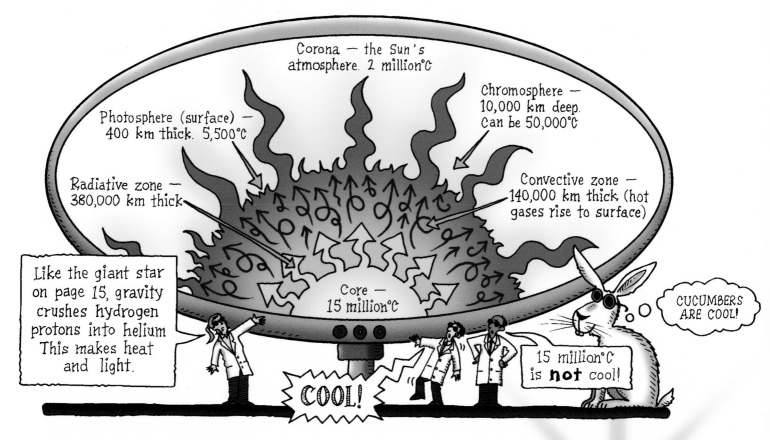

Corona — the Sun's atmosphere. 2 million°C

Chromosphere — 10,000 km deep. Can be 50,000°C

Photosphere (surface) — 400 km thick. 5,500°C

Radiative zone — 380,000 km thick.

Convective zone — 140,000 km thick (hot gases rise to surface)

Core — 15 million°C

Like the giant star on page 15, gravity crushes hydrogen protons into helium This makes heat and light.

COOL!

15 million°C is **not** cool!

CUCUMBERS ARE COOL!

SIX STUNNING SUN SECRETS

1 In the Sun's core, gravity crushes a square this size with a force of 700 million tonnes. Imagine 175 million hippos standing on each other's backs on top of that little square of YOUR BOOK!

2 Imagine you had a sunlamp made from a bit of the Sun the size of this square. The good news – it would give more light than 1,000 60-watt bulbs. The bad news – it would blast out ultraviolet rays, X-rays and gamma rays, and roast you to little crispy bits.

3 Oddly enough, the Sun's heat is a bit weedy. Gram for gram, your body actually makes more heat, and the only reason why the Sun is loads hotter than you is that it's loads bigger.

Mind you, the sun will be even more dangerous in five billion years time. By then all its hydrogen will have been used up. The core will collapse and heat up. KER-BAAM! The outer layers will blast into space and roast the Earth. After another billion years the core will blow and turn our planet into something that looks like one of your dad's cooking disasters.

Oh well, you've got plenty of time to finish this book!

BUT DON'T TAKE TOO LONG!

THE STUNNING
SOLAR SYSTEM HOLIDAY

So you need a holiday to get over all that sunburn? Well, here's an exciting idea – why not take a relaxing break on Mercury, Venus, the Moon or Mars? We've got the brochure, and all these places are ideal for that out-of-this-world vacation… OK, so I lied!

WANT A HORRIBLE TIME? WE CAN "PLANET"!

HORRIBLE HOLIDAYS PRESENT…

EEK!

The Mercury Month-off

- Guaranteed sizzling sunshine!
- Free helium in the atmosphere
- If you get too hot you can always chill out at night!

I FORGOT MY SUNCREAM!

The stunning small print
1. Mercury is 350°C. That's seven times hotter than the hottest Earth temperature. At night it's ten times colder than a freezer – Brrrr!
2. There's no air to breathe, and to fill one miserable Mickey Mouse-size balloon you'd need to suck in all the helium for 3.2 km around you.
3. Night and day each lasts 29 Earth days, so you've got plenty of time to roast or freeze to death (you choose).

The Venus Vacation

Enjoy a romantic break on Venus. It's sure to melt your heart (and the rest of you!)
- Warm welcome guaranteed!
- Stunning scenery
- Lots of atmosphere
- Exciting evening's entertainment.
You can watch the rocks glow!

I PREFER THE PRESSURE OF WORK

The stunning small print
1. Warm welcome … lots of atmosphere? Poohee! Venus offers an interesting choice of deaths. You could be…
- Boiled alive by the 470°C heat (Venus is the hottest planet).
- Crushed by the pressure of the atmosphere.
- Smothered by the carbon-dioxide gas that makes it up.
- Dissolved by the sulphuric acid clouds.
2. OK, so the scenery is stunning. One mountain, Maxwell Montes, is 2 km higher than Mount Everest and metal snow falls on its upper slopes.

WHOOPEE! NOW I ONLY WEIGH 19 STONE!

Moonlighting on the Moon

You'll be over the Moon when you stay on the Moon!
- Lose weight instantly! Weak gravity means you'll weigh only one-sixth of your Earth weight.
- Free moonlight!
- Lovely view of Earth will make you gasp!

The stunning small print
1. You'll gasp anyway – there's no air to breathe.
2. It's 110°C by day and -170°C at night.

Next out from Mars is the asteroid belt, but let's take a break from poring over holiday brochures and play an exciting computer game. It's a harmless bit of fun called…

THE EARTH-BLASTER MEGA-DEATH GAME

Your mission is to work out which of these eight asteroids is on course to hit Earth. WARNING – it could be more than one!

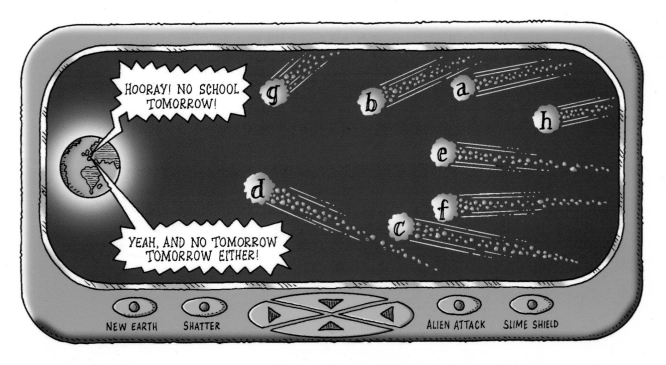

Answer:

a), c) and h) Don't be too scared of asteroids, readers. As you found out on page 77, space is huge so there's lots of room for them to miss us!

THE AWESOME OUTER PLANETS

After Mars, the planets are bigger and colder, and we couldn't find brochures for them because nobody wanted to go there. So instead, we decided to play a new quiz game…

Description A
This planet is mostly hydrogen and helium, and it's so light that it would float on water. There's no solid surface and the wind howls around it up to 1,770 km per hour. The planet is surrounded by rings made of rock and ice… Oops — what a giveaway!

Description B
This planet is rocky and smaller than Mercury, but it would still take a year to walk round. Mind you, there's not a lot to see and it's c-c-cold (about -200°C) so you'd best wear your scarf if you go there.

*WHAT'S THAT NOISE?

Description C

I'm talking about a big, bulging planet made of hydrogen gas. The planet spins at 45,500 km per hour — that's fast enough to stir up stunning storms in the planet's awful atmosphere. One huge hurricane looks like a big red spot.

Description D

This ghastly gassy planet is sure to give you the blues! In 1989 scientists spotted a storm raging around the planet at 1,000 km per hour. They called the storm "the scooter" — but I bet you'd need more than a scooter to escape from it!

Description E

The summer on this planet goes on for 21 years, and in all that time the Sun never sets — it seems to go up and down in the sky like a yo-yo! The bad news is the winter lasts 21 years too.

Answers:

1 E – Uranus. The yo-yo Sun and the long winters and summers are due to the way the planet tumbles through space.

2 A – Saturn. Easy, wasn't it?

3 D – Neptune

4 B – Pluto

5 C – Jupiter. Did you "spot" it?

Well, I guess we've come to the end of the Solar System, but there's just time to say a word of thanks to the first brainy boffin to peer at the Solar System through a telescope…

Brainy Boffin

Galileo Galilei (1564–1642)

As soon as Galileo peered at the night sky through the newly invented telescope he spotted spectacular space sights. He was the first to see Jupiter's four biggest moons, Saturn's rings and mountains on our moon.

Galileo wrote a best-selling book about his discoveries, but then everything went pear-shaped. He realized that the planets went round the Sun at a time when the Church taught that the planets and the Sun go round the Earth. When Galileo tried to write about his views, he was locked up for the rest of his life. Dangerous things, telescopes…

YOU THINK THE WHOLE UNIVERSE REVOLVES AROUND YOU!

ARE YOU CALLING ME A BIGHEAD?

But not half as dangerous as the unbelievable universe beyond our solar system. And that's where we're going next! But hold on, we've just received a signal from a distant part of the next chapter…

HELLO… WINGE, MOAN, GRUMBLE!

LET'S TAKE A LOOK AT...

THAT'S FAR OUT, MAN!

THE UNBELIEVABLE UNIVERSE

The size-sorted science in this fearsome final chapter is about as BIG as you can get. Prepare to be seriously stunned by the sheer stupendous size of the universe … and it's still getting bigger!

THE UNIVERSE – THE STORY SO FAR

A long, long time ago the universe was born. It grew and grew and grew. As it grew the universe cooled. After about 300,000 years electrons teamed up with protons to make atoms. Light shone between the atoms. The Big Bang was over – but it had left its mark. There was more matter and heat in some parts of the universe than others. Gravity began to pull this matter together to make the first stars. Scientists think this took about half a billion years…

PULL YOURSELF TOGETHER!

> ### Bet you never knew!
> Ever since the Big Bang the universe has been growing at almost the speed of light (300,000 km per second). And that means it's about 2 billion km larger than when you started this book!

Now read on, because it's time to meet a brilliantly brainy, big-headed boffin who was also a frightful fibber. You have been warned!

Brainy Boffin

Edwin Hubble (1889 –1953)

Edwin Hubble was strong and clever and good-looking. He was also fond of telling planet-sized porkie-pies such as…

I SAVED A MAN'S LIFE.

I WAS A WAR HERO.

I WAS A TOP LAWYER.

I KNOCKED OUT A CHAMPION BOXER.

SMALL MED BIG WHOPPER — LIE DETECTOR

BLAM!

WHOPPER

Even after Hubble became fantastically famous for his science discoveries, he went on telling fibs. And when the scientist died, his wife hid his body and wouldn't tell anyone where it was. Or was that another Hubble hoax?

Luckily when it came to science, Hubble really was telling the TRUTH…

MY TWO BIG DISCOVERIES by Edwin Hubble

1 I compared the brightness of stars. The dimmer stars were more distant and proved to be further away than anyone thought. The universe is BIGGER than we imagined!

2 I studied the colour of distant galaxies. They appeared red, owing to the stretching of light as the galaxies moved away. If the galaxies are moving away from us, the universe must be getting bigger!

The stunning size of the universe quiz

So could you be a brainy boffin like Hubble? Find out now by working out the size of the universe. To play, all you do is answer SMALLER, SPOT-ON or BIGGER to the following questions…

1 If the Earth and the Sun were 2.54 cm apart, the nearest star, Proxima Centauri, would be 64 km away.

2 Proxima Centauri is so far away that it would take 520 years to drive there in a car.

3 Our galaxy, the Milky Way, has so many stars that it would take 60 years to count them all.

4 The Milky Way is so gigantic that it would take 200 years to send a signal to an alien planet on the other side of the galaxy and get a reply.

A **light year** is the distance light travels in a year.
One light year = 9.46 million million km, give or take the odd metre.

A **galaxy** is a giant group of stars.

Answers:

1 BIGGER – it would be 6,436 km away.

2 BIGGER – it would actually take 52 million years at a steady 88.5 km per hour. Let's hope there are space toilets on the way.

3 BIGGER – it would take 6,000 years to count all 200,000 billion of them.

4 BIGGER – the Milky Way is 100,000 light years across. Assuming you used a laser, you might get a reply in 200,000 years – that's if there were aliens on your target planet.

So the universe is a bit on the big side. But that's just for starters – in fact there are over 140 billion galaxies out there and many of them are BIGGER than the Milky Way… OK, you can stop gasping now.

HOW TO BE A STAR ASTRONAUT IN FIVE EASY LESSONS

So you want to explore the breathtakingly BIG universe? Great – you've come to the right page!!! Why not join the shrinking scientists on our astronaut training course?

THE HORRIBLE SCIENCE ASTRONAUT TRAINING COURSE

Lesson 1 – Driving a rocket ain't rocket science

All you do is switch on the engine…

Hot gases blast out this end

X-ray view

BLAM!

BOOSH!

Hot gases blast against inner walls of rocket

Rocket is pushed forward

EEK! WHERE'S THE STEERING WHEEL?

Your rocket doesn't have brakes or a steering wheel. To change your course, you fire another rocket engine in the opposite direction to the one you want to travel.

Lesson 2 – Being weightless

Away from Earth's gravity, your body is weightless. And that means everything in your body, including your half-digested sick and blood, will be weightless too. You may notice some unusual changes as your body fluids float upwards…

Legs and waist shrink

Nose feels blocked

Puffy eyes

Fluids float to kidneys, causing an urge to wee

Lesson 3 – Take a breather

There's no air in space. This can be a problem, since your spacecraft can get very smelly – especially if a fellow astronaut has been scoffing beans. If your head hurts, and you gasp and feel sick, this could be due to the smell – or it could be due to a dangerous build-up of carbon-dioxide gas breathed out by your body. Use lithium salt to soak up the carbon dioxide – but remember, there's one thing you should NEVER do…

I'LL JUST OPEN THE WINDOW!

Lesson 4 – Living with weightlessness

Being weightless has good and bad sides…
• GOOD SIDE You can perform superb somersaults and awesome acrobatics, and play weird games like upside-down, slow motion darts.
• BAD SIDE Your brain gets confused about which way up you should be. This causes dizziness, sweating and throwing up.
• It's so easy to float around that your muscles and bones don't have much work to do. They start to waste away and you need to exercise for two hours a day to keep them strong.
• Anything you spill forms round globules and floats around your spacecraft. An accident on the space toilet doesn't bear thinking about.

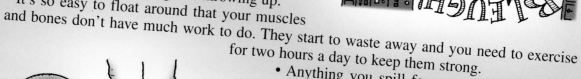

WHAAAA! BLEUGH!

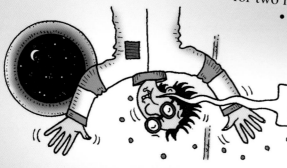

IT'S FUN CATCHING THESE CHOCOLATE PEANUTS!

Lesson 5 – Coming back is hard to do

Sooner or later you'll want to go home – after all, there's only so much floating around in a metal box that you can take. But the return to Earth is actually the most dangerous part of your mission… When your spacecraft hits the atmosphere, it's moving so fast that air molecules can't get out of the way. They're shoved together in front of the craft and heat up by rubbing (friction). Your craft heats up too, and it could burn up. Hopefully your craft's heat shields will protect you – but if not, your goose is cooked, and that goes for the rest of you! Still wanna be an astronaut?

Bet you never knew!

In 1963 astronaut Gordon Cooper circled the Earth 22 times in his spacecraft. Part of his mission was to take a sample of his wee for testing. Unfortunately, the wee escaped and floated around inside his smelly spacecraft.

NO WAY!

WELCOME TO THE HORRIBLE SCIENCE
STAR-STRUCK STAR-SPOTTER'S
COMPETITION!

Here's your chance to prove you know one end of a telescope from the other!

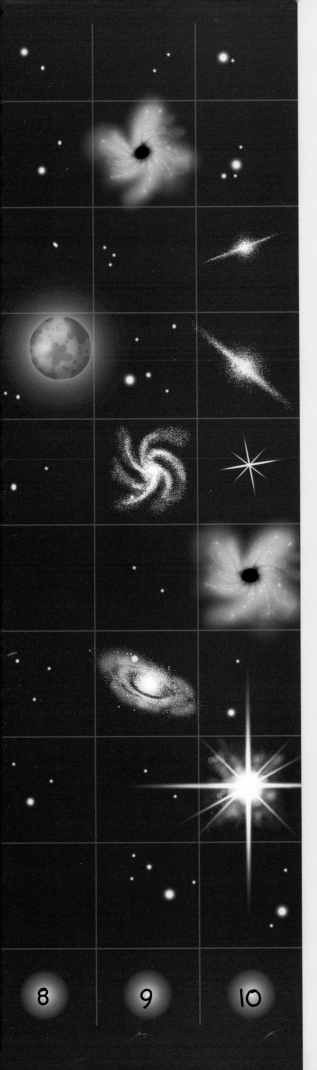

How to play

1 Study the Space-spotter's guide below.
2 Spot the objects listed and note down their co-ordinates in the star grid
– for example, there's a red giant star in square A1.
WARNING: THERE MAY BE MORE THAN ONE OF EACH OBJECT!
3 Check your answers and GOOD LUCK!

SPACE-SPOTTER'S GUIDE

1 Red giants
No, we're not in the land of fairy tales. A red giant is what happens when a star like the Sun runs out of hydrogen (see page 79 for a quick reminder).

2 White dwarf
What's left after a red giant's blown itself out. That's how our poor old Sun's going to end up.

3 Red dwarf
A dim star one-tenth the size of the Sun. Proxima Centauri is one of these – which is why it's not worth a 52-million-year drive to get there.

4 Brown dwarf
A wannabe star that never got big enough to shine. It probably sits around all day feeling sorry for itself.

5 Supernova
A giant exploding star that goes pop with the force of 1,000 billion atom bombs. If one went off closer than 33 light years away we'd get roasted by X-rays and gamma rays and look like the mummy on page 19. DON'T PANIC READERS! We're not that close to one of these nasty blasters!

6 Neutron star
After the blast what's left of the giant star shrinks to the size of a large city under its own gravity. The gigantic gravity of a neutron star means the slimy protist from page 29 would weigh as much as two ocean liners. Imagine that in your mouth!

7 Black hole
A star with ten times more mass than the Sun doesn't end up as a neutron star. Its giant gravity sucks a hole in the universe. This horrible hole is black because not even speedy light can escape. So everything gets sucked in – planets, stars, moons, asteroids and the astronauts' baked beans.

8 Quasar (Kway-zar)
Galaxies form around huge black holes that slurp in stars like giant plugholes. As matter whirls into the black hole it heats up and vast explosions result. This is a quasar, but your legs needn't turn to jelly – the nearest quasars are billions of light years away.

9 Galaxy
Some galaxies are shaped like spirals and some look like alien flying saucers. How many of each can you spot?

10 The Milky Way
Our friendly local galaxy is spiral-shaped, but the area where there's the most stars looks like a giant's dribble to us because we see it from the side. We're about 25,000 light years from the black hole that probably lurks in the centre.

11 Andromeda galaxy
This spiral galaxy is two billion light years off, but it's heading our way. In the time it's taken you to read about it, it's whizzed 1,000 km closer. DON'T SCREAM! It won't hit us for five billion years and when it does, most of the stars will pass each other (space is big enough to avoid collisions).

12 Alien planets
Planets may be common in the Milky Way but there's no proof that aliens live on them. According to one guess, planets with intelligent life on them could be 200 light years apart, so the aliens probably won't be dropping in for coffee… By the way, can you spot the purple slobwobbler in the vastness of space?

Answers
1 A1, G7
2 B7, C3, E7
3 A2, C4, D2, G1
4 F2
5 B10
6 A4, E10
7 C1, D10, E4, H9
8 D3
9 A3, E9, F4, F10, G3, G10
10 The band running from A5 to G6 – there are three of each type.
11 C9
12 A6, E1, F8. The purple slobwobbler is in B2

THE END OF EVERYTHING

And now for a terrible time trip without a toilet stop… For their final mission the shrinking scientists are going visit the universe 100,000 billion years in the future. And by then it'll be so gobsmackingly GIGANTIC that I can't even begin to tell you how big it is.

The shrinking scientists in … the spaced-out future

Well, here we are in the future, but where are the stars? It's very dark and very cold. The universe is so BIG that gravity can't pull matter together to make new stars. In fact, most matter has already been guzzled by black holes.

So the outlook of the universe is dark and dismal. In this far-off frozen future the only light will come from dim dwarf stars slowly sputtering out like burnt cinders. Just imagine a power cut that lasts forever. What's that? You want to stock up on torch batteries? Cheer up, it won't happen just yet…

STOP PRESS
We interrupt this book for some important news about the future of the universe. IT'S OFFICIAL – Over the past few billion years the universe has been growing even faster!

The **COSMIC CHRONICLE** — 1998 —

– WIN A PLANET!

UNIVERSE GROWS EVEN FASTER!

ALSO IN THIS WEEK'S ISSUE…

THE UNIVERSE – IS BIGGER BETTER?

ALIEN TELLY GUIDE

NOW BIGGER THAN EVER!

Scientists reckon the cosmos is growing faster. It's all due to a mysterious force called dark energy pressing on the fabric of space. We asked a scientist about dark energy but after writing lots of hard sums on a blackboard he admitted scientists were in the dark.

The editor writes…
Oh well – we could do with a bit more space. Now where did I leave my spacecraft keys?

Hmmm – I wonder what life will be like in the distant future? To find out I've just invented the sort of creature we could evolve into. Meet Tinpot the robot…

THE HORRIBLE SCIENCE INTERVIEW

Tinpot's tale

Tinpot is adapted to life in a universe with almost no energy.

CREAK!

HELLO, SIGH!

Tinpot plugs into giant solar panels to collect ever-decreasing starlight. He makes food from light like a plant.

Slow movements to save scarce energy

Horrible Science: So what's life like in the far future?

(100-year delay while Tinpot gets enough energy to reply.)

Tinpot: Er — what was the question again? Oh yes, I remember… it's unbearably boring. It's cold, it's dark and nothing much happens. I've got a good mind to switch off my power cell. The most exciting thing that's happened to me is a touch of rust, but that was over a million years ago — sigh!

Horrible Science: Why don't you read a Horrible Science book and lighten up?

Of course we can't be sure that humans will evolve into Tinpot. And we can't be sure what the future holds in our ever-growing universe. But you can be sure that scientists won't stop trying to find out. After all, scientists want to find out about everything. Right now they're gazing at galaxies, stars and black holes and searching for new planets and alien life. And they're digging for dinosaur bones and trying to uncover the secrets of the human body, microbes, molecules, atoms and the very instant of the Big Bang. And the answers are out there … er, somewhere!

EPILOGUE:
THAT'S THE SIZE OF IT!

So you've finished *The Stunning Science of Everything*? And you've read *every* word on every page? CONGRATULATIONS – you definitely deserve this certificate!

It's been a long journey from the tiny big bang through ever more sizeable science to the giant goodbye of the enormous universe. But now we can relax with a bottle of fizzy pop and enjoy "THE BIG PICTURE OF EVERYTHING" on the next page. This masterpiece by our Horrible Science artist sums up the whole of stunning science. Thanks a million, Tony!

Wow – look at all that science! But there's something more. Can you see how everything is part of something larger?

• Atoms can be part of molecules.

• Molecules make up microbes and cells.

• Cells make up YOU!

• You and me and all the other animals and bugs and plants make up the teeming life of Planet Earth.

• Earth and all the other planets and stars are part of the unbelievably BIG universe…

And there's more. It turns out that the little things are vital to the big things…

• If it wasn't for atoms and quarks, you and me and everything else in the universe wouldn't exist.

• Stars can't shine without atoms to crush and light to blast.

• Without tiny electrons to sense with its snotty nose, the echidna couldn't find any beetle grubs for breakfast.

• Plant-eating animals ought to say "THANKS GUYS!" to the bacteria that help them digest their food.

• Even insects have their uses. Where would buffalo be without beetles to bury their dung? Knee-deep in poo – that's where!

• DNA tells your body how to make a new you.

• And on the bad side – microbes can kill people.

Did you spot all the Francium atoms? They appeared on the following pages: 18, 20, 27, 34, 38, 42, 47, 50, 54, 58, 61, 63, 68, 72, 74, 77, 82, 85, 90.

There's a moral and a message here, and it's this… No matter how tiny and timid something is, no matter how weak and weedy – it's still vital. If you look closely, everything is connected to make one big, amazing, ever-growing universe. And that's the stunning truth!

STUNNING TIMELINE

About 13.7 billion years ago
The Big Bang kick-starts the universe and the most exciting stuff happens in the first second. After that it's downhill all the way.
Ten minutes later
All the matter in the universe has formed. If you want to know what's the matter, here's your answer!

BLAM!

About 13.4 billion years ago
All the hydrogen atoms and most of the helium has formed. Life was such a gas back then!

FSSSS!

IT'S SOMETHING TO GAS ABOUT

4.6 billion years ago
The Sun and the planets form from gas and dust. Luckily for us, it all came together in the end.

YEAH, PHEW!

Around 12.5 billion years ago
The first galaxies form in the new-look spaced-out universe. Quasars blast out the hot hearts of the new galaxies. It was a crazy queasy quasar time…

VooOSH!

4 billion years ago
Earth gets clunked by a planet the size of Mars. A big splurge of red-hot rock races off into space and turns into the Moon. It was the biggest rock and roll hit ever!

OOOF!

CLONK!

100,000 years ago
Several species later, the upright apes evolve into very wise humans (which is odd, since some of us don't act too wisely). Oh well, here are some stunningly wise scientists.

1609 – Italian scientist
Galileo Galilei is the first person to spy on the Solar System through a telescope.

1687
Isaac Newton publishes his work on gravity.

1858
Charles Darwin announces his ideas about evolution.

1905 and 1915
Albert Einstein makes sense of time and space and energy and matter. What a brilliant brainbox!

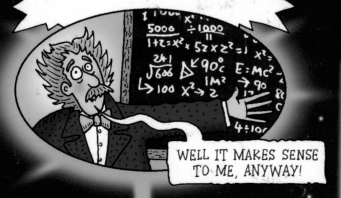

1923 and 1929
Edwin Hubble gets big ideas about the universe.

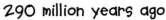

290 million years ago
Ruthless reptiles take over the world. They were an ugly lot and some of them were hairy. (They were your ancestors.)

230 million years ago
The dreaded dinosaurs dine off the reptiles until most of them die off. The first mammals go into hiding.

65 million years ago
D-day for dinosaurs. D = Disaster, Destruction and Don't-go-out-without-a-crash-helmet Day. Yes, it's die-now-dinos day.

55 million years ago
After millions of years of small-time life, the mammals come into their own. Welcome to the age of BIG HAIR.

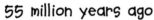

7 million years ago
An upright ape in Africa starts to get big ideas. Say Hi! to Granny…

1.64 million years ago
The Ice Age era starts and it's still with us. It must have been chilly before clothes were invented…

4 billion years ago
The first slimy life forms appear in the sea. I guess they were just a drop in the ocean.

750 million years ago
Earth turns into a snowball – or did it? Well, it's a chilling thought!

520 million years ago
Eel-like creatures are the first life forms with backbones. I bet they got backache and felt eel, I mean ill.

450 million years ago
Plants and mites invade the land. And to this day no one knows how to get rid of the mites.

370 million years ago
Fish crawl on to land and evolve into amphibians. Suddenly the world is ruled by giant frogs with attitude.

350 million years ago
Giant millipedes crawl about. Some of them evolved to be 2 metres long.